MW01113977

*Environmental Science and Engineering*
*Subseries: Environmental Science*

Series Editors: R. Allan • U. Förstner • W. Salomons

For further volumes:
http://www.springer.com/series/7487

Md. Danesh Miah · Man Yong Shin · Masao Koike

# Forests to Climate Change Mitigation

## Clean Development Mechanism in Bangladesh

Md. Danesh Miah
Shinshu University
Forest Policy Laboratory
Faculty of Agriculture
8304 Minamiminowa-Mura
Kami Ina Gun
399-4598 Nagano-Ken
Japan
dansforestry@yahoo.com

Prof. Dr. Masao Koike
Shinshu University
Forest Policy Laboratory
8304 Minamiminowa-Mura
Kami Ina Gun
399-4598 Nagano-ken
Japan
makoike@shinshu-u.ac.jp

Prof. Dr. Man Yong Shin
Kookmin University
Department of Forest Science
136-702 Seoul
Korea, Republic of (South Korea)
yong@kookmin.ac.kr

ISSN 1863-5520
ISBN 978-3-642-13252-0 e-ISBN 978-3-642-13253-7
DOI 10.1007/978-3-642-13253-7
Springer Heidelberg Dordrecht London New York

Library of Congress Control Number: 2010930774

© Springer-Verlag Berlin Heidelberg 2011
This work is subject to copyright. All rights are reserved, whether the whole or part of the material is concerned, specifically the rights of translation, reprinting, reuse of illustrations, recitation, broadcasting, reproduction on microfilm or in any other way, and storage in data banks. Duplication of this publication or parts thereof is permitted only under the provisions of the German Copyright Law of September 9, 1965, in its current version, and permission for use must always be obtained from Springer. Violations are liable to prosecution under the German Copyright Law.
The use of general descriptive names, registered names, trademarks, etc. in this publication does not imply, even in the absence of a specific statement, that such names are exempt from the relevant protective laws and regulations and therefore free for general use.

*Cover design*: Integra Software Services Pvt. Ltd., Pondicherry

Printed on acid-free paper

Springer is part of Springer Science+Business Media (www.springer.com)

# Abstract

Accurate accounting of carbon stocks and stock changes in forest ecosystem is necessary for the improved greenhouse gas inventory which was made mandatory by the United Nations Framework Convention on Climate Change and its Kyoto Protocol. The Kyoto Protocol provides for the involvement of developing countries in an atmospheric greenhouse gas reduction regime under its Clean Development Mechanism (CDM). Bangladesh, a densely populated subtropical country in South Asia, has huge degraded forestlands which can be reforested by the CDM projects. To visualize the potential of the forestry sector in developing countries for emission mitigation, carbon sequestration potential of different species in different types of plantations should be integrated with the carbon trading system under the CDM of the Kyoto Protocol. Fossil fuel substitution by biomass fuel and its efficiency can also be important options for the Clean Development Mechanism (CDM) projects in Bangladesh.

The book finds that afforestation and reforestation (A/R) can be one of the greatest choices in mitigating global warming by increasing the carbon sink in Bangladesh under the CDM. Avoiding deforestation also can be a great option by decreasing the carbon sources in Bangladesh, but this is not recognized by the CDM yet. The study shows that the greatest reforestation success of the Republic of Korea can be a better lesson for Bangladesh to increase the carbon sink in the forests. It confirms that bioenergy projects are attractive and CDM provides complementary options for international cooperation toward sustainable development in Bangladesh. The study shows that burning of biomass in the traditional cooking stove in Bangladesh has a severe implication on the deforestation and greenhouse gas emission to the atmosphere. It also confirms that innovation of the improved cooking stove can be critical to the involvement of the CDM. The results show that tree tissue in the forests of Bangladesh stores 92 tC $ha^{-1}$, on average. The results also reveal a gross stock of 190 tC $ha^{-1}$ in the plantations of 13 tree species, ranging in age from 6 to 23 years. The study confirms the huge atmospheric $CO_2$ offset by the forests if the degraded forestlands are reforested by the CDM projects, indicating the potential of Bangladesh to participate in carbon trading for both its economic and environment benefits.

The book suggests the capacity building and policy changes in Bangladesh to comply with the CDM modalities. It also suggests Bangladesh to learn the reforestation success from the Republic of Korea. Research and development (R&D) of efficient biomass burning has been recommended for Bangladesh.

The outcome of this book will be of great importance to both national and international policy makers in the field of global warming mitigation. The energy policy makers, administrators, and the international CDM investors will find this study critical to the national sustainable development in developing countries like Bangladesh.

# Contents

# List of Abbreviations

| | |
|---|---|
| ADB | Asian Development Bank |
| AIJ | Activities Implemented Jointly |
| A/R | Afforestation/Reforestation |
| B | Billion |
| BB | The Bangladesh Bank |
| BBS | Bangladesh Bureau of Statistics |
| BPDB | Bangladesh Power Development Board |
| BCSIR | Bangladesh Council of Scientific and Industrial Research |
| CBD | Convention on Biological Diversity |
| CDM | Clean Development Mechanism |
| CER | Certified Emission Reduction |
| $CH_4$ | Methane |
| CIDA | Canadian Institute of Development Agency |
| cm | Centimeter |
| CO | Carbon monoxide |
| COP | Conference of the Parties to the United Nations Framework Convention on Climate Change |
| dbh | Diameter at Breast Height |
| DC | District Commissioner |
| dm | Dry matter |
| ESSD | Environmentally and Socially Sustainable Development |
| FAO | Food and Agriculture Organization |
| FMP | Forestry Master Plan |
| FRA | Forest Resource Assessment |
| g | Gram |
| GDP | Gross Domestic Product |
| GHG | Greenhouse gas |
| GOB | Government of Bangladesh |
| GoRok | Government of the Republic of Korea |
| GtC | Giga tonne Carbon |
| ha | Hectare |
| IFRD | Institute of Fuel Research and Development |
| IPCC | Intergovernmental Panel on Climate Change |

| | |
|---|---|
| JI | Joint Implementation (Outlined in Article 6 of the Kyoto Protocol) |
| KFS | Korea Forest Service |
| kg | Kilogram |
| kgoe | Kilogram of oil equivalent |
| $Km^2$ | Square kilometer |
| LULUCF | Land Use, Land-Use Change and Forestry |
| m | Million |
| $m^2$ | Square meter |
| $m^3$ | Cubic meter |
| MAI | Mean Annual Increment |
| MoEF | Ministry of Environment and Forests (in Bangladesh) |
| MoEROK | Ministry of Environment of the Republic of Korea |
| MOP | Meeting of the Parties to the Kyoto Protocol |
| NTFP | Non-Timber Forest Products |
| NGO | Non-Governmental Organization |
| OECD | Organization for Economic Co-operation and Development |
| ppm | Parts per million |
| R&D | Research and Development |
| ºC | Degree Celsius |
| SBSTA | Subsidiary Body for Scientific and Technological Advice |
| SEHD | Society for Environment and Human Development |
| SESRTCIC | Statistical, Economic and Social Research and Training Centre for Islamic Countries |
| SSS | Soil Survey Staff |
| t | Tonne |
| tC | Tonne Carbon |
| UNEP | United Nations Environment Program |
| UNFCCC | United Nations Framework Convention on Climate Change |
| USCB | U.S. Bureau of the Census |
| WB | The World Bank |
| WEC | World Energy Council |
| yr | Year |

# List of Figures

# List of Tables

# Chapter 1
# Climate Change Mitigation by the Forestry Options in Bangladesh – A Synthesis

**Abstract** Understanding the background of climate change along with its causes, consequences, and responses of the global community is important for climate change mitigation. This chapter attempts to construct a theoretical framework on how climate change disrupts the human and ecological systems and how forest land use mitigates this. As a global response to the climate change, Kyoto Protocol and its Clean Development Mechanism (CDM) regarding forestry and bioenergy promotion have been importantly discussed here. This chapter finds a severe consequence of climate change on the globe, tropical developing countries in particular. It argues a strong impact of forest-based CDM on the climate change mitigation as well as sustainable development in the non-Annex I countries. This chapter also describes the scope, objectives, and general methodological approach of the book.

## 1.1 General Introduction

Climate change is a hotcake issue in the present day, especially due to the accumulation of greenhouse gases (GHG), principally carbon dioxide ($CO_2$), in the atmosphere because of emissions caused by industrial activities and combustion of fossil fuels for non-industrial activities and deforestation and other land-use changes (Fearnside 2006; Houghton 2005; Nordell 2003). Mayaux et al. (2005) and Achard et al. (2002) have reported that the world's humid tropical forests had been disappearing at a rate of about 5.8 ($\pm$1.4) m ha $yr^{-1}$, with a further 2.3 ($\pm$0.7) m ha $yr^{-1}$ of forests visibly degraded between 1990 and 1997. Natural and man-made disasters, which have often been front-page news in the last couple of years, maintained their high frequency. Over the last decade (1990 s), both the numbers and severity of the weather disaster events increased. The number of events and economic losses were three times bigger in comparison to the 1960 s (Mirza 2003). These catastrophes have imposed severe pressure on poor economies, shattered infrastructure, and made the poor more vulnerable. It has been estimated that people in low-income countries are more likely to die in natural disasters than people in high-income countries (Alcántara-Ayala 2002; Mirza 2003).

Md.D. Miah et al., *Forests to Climate Change Mitigation*, Environmental Science and Engineering, DOI 10.1007/978-3-642-13253-7_1, © Springer-Verlag Berlin Heidelberg 2011

Kram et al. (2000) point out that the distribution of both income and GHG emissions is very unbalanced between various world regions. The relative importance of individual gases and sources of emissions differs from region to region. Kram et al. (2000) analyzed that currently developing countries account for about 46% of all emissions, but by 2100 no less they would contribute 67–76% of the global total, while the total income generated in these countries would reach 58–71% from only 16% in 1990. But Kram et al. (2000) conclude that when population size and the levels of affluence in the developing countries confront with the potential severity of climate change-induced damages, the scenarios are very different. Higher population densities and lower income make the countries more vulnerable for adverse climate change impacts, and that lower income creates less favorable conditions for mitigation and/or adaptation measures. Domestic activity in developed countries must therefore receive priority, and the main emitters of $CO_2$ should assume their responsibility for tackling the causes. In the Protocol, which was adopted in Kyoto, Japan, in 1997, industrialized countries committed themselves to reduce their combined GHG emissions by more than 5% relative to the level in 1990, in the period between 2008 and 2012. The Kyoto Protocol recognizes forestry and land-use change activities as sinks and sources for atmospheric carbon. In a special report on land use, land-use change, and forestry, the Intergovernmental Panel on Climate Change (IPCC) concludes that activities in the realm of land-use change and forestry provide an opportunity to affect the carbon cycle positively (IPCC 2000). FAO (2001) proposes three possible strategies for the management of forest carbon. The first is to increase the amount or rate of carbon accumulation by creating or enhancing carbon sinks (carbon sequestration). The second is to prevent or reduce the rate of release of carbon already fixed in existing carbon sinks (carbon conservation). The third strategy is to reduce the demand for fossil fuels by increasing the use of wood, either for durable wood products (i.e., substitution of energy-intensive materials such as steel and concrete) or for biofuel (carbon substitution). A number of carbon sequestration and carbon conservation initiatives have already been developed, including Activities Implemented Jointly (AIJ) under the UNFCCC and land-use change and forestry carbon projects (FAO 2001).

Knowledge of the quantity, distribution, and partitioning of carbon and any change, which takes place over time, can contribute greatly to quantify the carbon sequestration by forests, which is thought to be a promising means of reducing atmospheric $CO_2$, an important GHG. Carbon sequestered by the national forests and afforestation/reforestation (A/R) projects was well identified by Hansen et al. (2004), Cannell (2003), Pussinen et al. (1997), Karjalainen (1996), Ravindranath & Somashekhar (1995), Ismail (1995), etc. The carbon sequestration potential of forests is specific to the species, site, and management involved, and it is therefore very variable. Assuming a global land availability of 345 m ha for A/R and agroforestry activities, Brown et al. (1996) estimate that approximately 38 Gt of carbon could be sequestered over the next 50 years, i.e., 31 Gt by A/R and 7 Gt through the increased adoption of agroforestry practices.

The Republic of Korea is a typical mountainous country where forest cover is about 65% of the total landmass covering about 6.4 m ha (Brown 2006; KFS 2006).

During the Japanese Occupation (1910–1945) and the Korean War (1950–1953) in the south part of the Korean Peninsula, illegal logging for forestry utilization resulted in a remarkable decline in the average stock volume per hectare from some 100 $m^3$ in early 1900 to 10 $m^3$ in 1960 (Lee 2003). To rehabilitate the degraded forestland, the Korean Government undertook a series of 10-year forest development plans including the massive reforestation programs. In 1999, a total of 97% of the degraded forestlands were rehabilitated very successfully, by planting about 12 billion trees and raising growing stock to 76.4 $m^3$ $ha^{-1}$ in 2004 (KFS 2006). Brown (2006) reports that the Republic of Korea 'is in many ways a reforestation model for the rest of the world.'

The implication of the legal frameworks of the Clean Development Mechanism (CDM) on the Kyoto Protocol is important for creating CDM forests in Bangladesh and to achieve the 'Certified Emission Reduction' (CER). In addition to this, appropriate policy formulation is also critical to achieve the carbon credits in the country. Finding out the forestry options for mitigating global warming is critical to the reorientation of Bangladesh to participate in the international global warming mitigation initiatives. But these options were not analyzed comprehensively in the past. Traditional cooking stoves in Bangladesh usually burn woody biomass, which is usually inefficient, emitting a lot of GHGs to the atmosphere and overutilizing the forests. Introducing an improved cooking stove is expected to contribute to global warming mitigation. The study aims at quantifying potential carbon sequestration and production of forest biomass in Bangladesh. The study also aims at implicating the traditional cooking stoves and its improvement to fit with the CDM. In addition to this, the study discusses the reforestation success of the Republic of Korea which can be a lesson to Bangladesh. It also discusses the policy issues to expedite the development of the CDM forests. The findings of this study would be useful for the policy makers, environmentalists, and investors to the CDM forests.

## 1.2 Theoretical Framework

### *1.2.1 Climate Change Disrupts the Human and Ecological Systems and Forest Land Use Mitigates This*

The industrial revolution made major changes in the technology, socio-economy, and cultures in the late 18th and early 19th centuries that began in Britain and spread throughout the world. The massive industrialization replaced the manual labors at that time. The technology-dominated economy was mostly dependent on the energy produced from the fossil fuel which keeps pace even now. It has been confirmed that fossil fuel burning has increased the GHGs to the atmosphere creating global warming. Among the GHGs, the concentration of $CO_2$ has been confirmed as the largest. IPCC (2001a) reported that the atmospheric concentration of $CO_2$ has increased by 31% since 1750. It has also been espoused by Quay (2002). IPCC (2001b) estimated that about three-quarters of the anthropogenic emissions of $CO_2$ to the atmosphere during the past 20 years is due to fossil fuel burning. The rest is predominantly due to land-use change, especially deforestation. Before the industrial revolution, the concentration of $CO_2$ was about 280 parts per million (ppm) for thousands of years until about 1800 (IPCC 2001b). The revolution increased that concentration from 368 ppm in 2000 (IPCC 2001b) to 380 ppm in 2004 (Takahashi 2004). The rate of increase of atmospheric $CO_2$ concentration has been about 1.5 ppm (0.4%) per year over the past two decades (IPCC 2001a). Due to the increased atmospheric concentration of $CO_2$, the global average surface temperature has also increased since 1861. Over the 20th century, the increase has been $0.6 \pm 0.2$°C (IPCC 2001a). Satellite data show that there are very likely (90–99% chance) to have been decreases of about 10% in the extent of snow cover since the late 1960 s (IPCC 2001a). Due to this decrease of snow cover, the sea level also rose between 0.1 and 0.2 m during the 20th century which has been proven by tide gauge data (IPCC 2001a). The precipitation has increased by 0.5–1% per decade in the 20th century over most mid- and high latitudes of the Northern Hemisphere continents, and the rainfall has also increased by 0.2–0.3% per decade over the tropical (10–30°N) land areas (IPCC 2001a). It has been proven by facts that all the events have resulted in climate change (Fig. 1.1). However, there is strong evidence that human activities have influenced the climate change. Salam and Noguchi (2005) have analyzed for 113 countries (19 developed and 94 developing) that combustion of fossil fuels

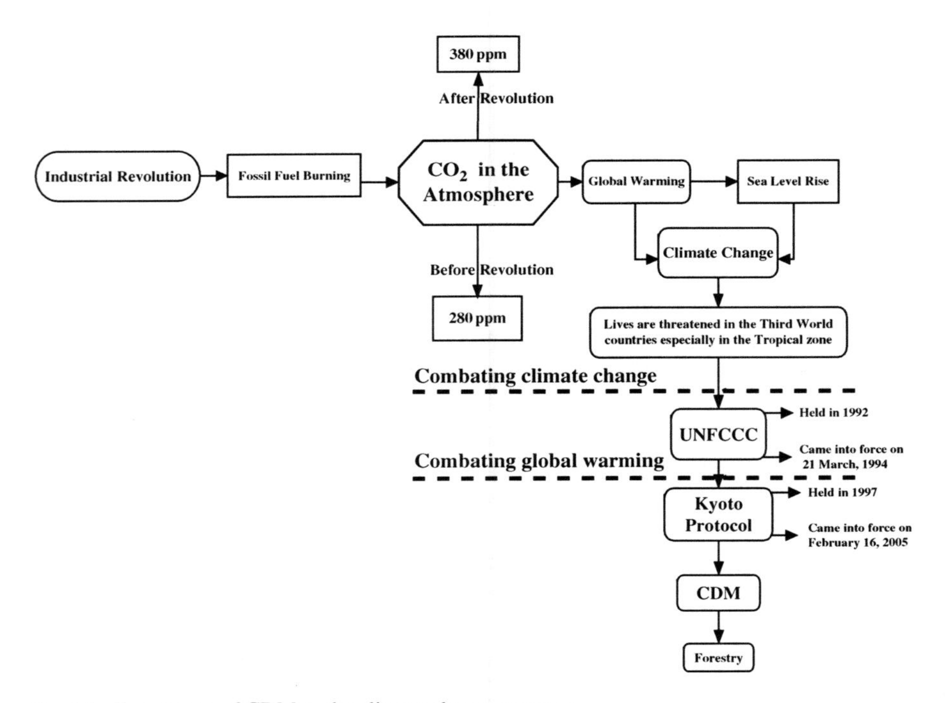


**Fig. 1.1** Emergence of CDM to the climate change response

has significant positive influence on $CO_2$ emissions in both developed and developing countries. Terrestrial ecosystem strength has a significant negative influence on $CO_2$ emissions. den Elzen et al. (2005) found that the relative contributions of different nations to global climate change from emissions of GHG alone are quite robust, despite the varying model complexity and differences in calculated absolute changes. For the default calculations, the average calculated contributions to the global mean surface temperature increase in 2000 are about 40% from the Organization for Economic Co-operation and Development (OECD), 14% from Eastern Europe and the former Soviet Union, 24% from Asia, and 22% from Africa and Latin America. Recent studies by Stebich et al. (2005), Berglund (2003), and Vincens et al. (2003) have also confirmed the human impact on climate change. Different climate models indicate that future human activities also will increase the concentrations of GHGs to the atmosphere which will cause further climate changes which could have large negative impacts on human and ecological systems (IPCC 2001c; Johns et al. 2003). IPCC (2001a) reports that emissions of $CO_2$ due to fossil fuel burning are virtually certain (greater than 99% chance) to be the dominant influence on the trends in atmospheric $CO_2$ concentration during the 21st century. The net effect of land and ocean climate feedbacks as indicated by the models is to further increase projected atmospheric $CO_2$ concentrations by reducing both the ocean and land uptake of $CO_2$. By 2100, carbon cycle models project atmospheric $CO_2$ concentrations of 540–970 ppm. The globally averaged surface temperature is projected to increase by 1.4–5.8°C over the period 1990–2100. Global mean sea level is projected to rise by 0.09–0.88 m between 1990 and 2100. This is due primarily to thermal expansion and loss of mass from glaciers and ice caps.

The projected climate change will hamper the lives drastically, especially in the low-lying, coastal developing countries if the world fails to curtail its $CO_2$ emissions. About half of the world's population lives in coastal zones, although there is a large variation among countries. Changes in climate change will affect coastal systems through sea-level rise and an increase in storm surge hazards and possible changes in the frequency and/or intensities of extreme events.

To combat the climate change, international negotiations led to a first step through adopting United Nations Framework Convention on Climate Change (UNFCCC) in 1992 which came into force in 1994. In the UNFCCC, there were no compulsory commitments of the Parties.

So, to address the more specific cause of the climate change and set the commitments of the Parties, another Protocol was adopted in 1997 at the COP3, which came into force in 2005. Through Article 12 of the Kyoto Protocol, one flexible mechanism, Clean Development Mechanism (CDM), was created to fulfill the reduction commitment of the Annex I countries buying carbon credits from the developing countries (Non-Annex I).

It has been recognized that terrestrial ecosystem plays an important role in global carbon sequestration. Carbon sequestration by changes in the use and management of forests can make a meaningful contribution to reducing atmospheric $CO_2$ (IPCC 2001b). Niles et al. (2002) analyze that over the next 10 years, 48 major tropical and subtropical developing countries have the potential to reduce the atmospheric carbon burden by about 2.3 b tC. So, forestry sector has been importantly included in the CDM for creating carbon credits in the Non-Annex I countries. The effect of industrial revolution on the atmospheric concentration of GHG, global warming, and climate change through the adoption of CDM for forestry has been shown in Fig. 1.1 as a schematic diagram.

## *1.2.2 Kyoto Protocol Obliges the GHG Reduction Commitment to the Annex I Parties*

To effectively address the anthropogenic climate change, GHG emission was recognized as important by the Kyoto Protocol. It was the product of the COP3 of the UNFCCC with the target of 'stabilization of GHG concentrations in the atmosphere at a level that would prevent dangerous anthropogenic interference with the climate system.' The Protocol sets specific reduction targets and timetables for reducing net GHG emissions from the Annex I (industrialized) countries. It calls for Parties to reduce their annual emissions by 5% below 1990 emissions. The Protocol with the ratification of the Russian Federation came into force on February 16, 2005. As of September 28, 2006, 166 states and regional economic integration organizations have deposited instruments of ratifications, accessions, approvals, or acceptances (Fig. 1.2). A total of 37 Annex I and 129 Non-Annex I Parties compose this Kyoto

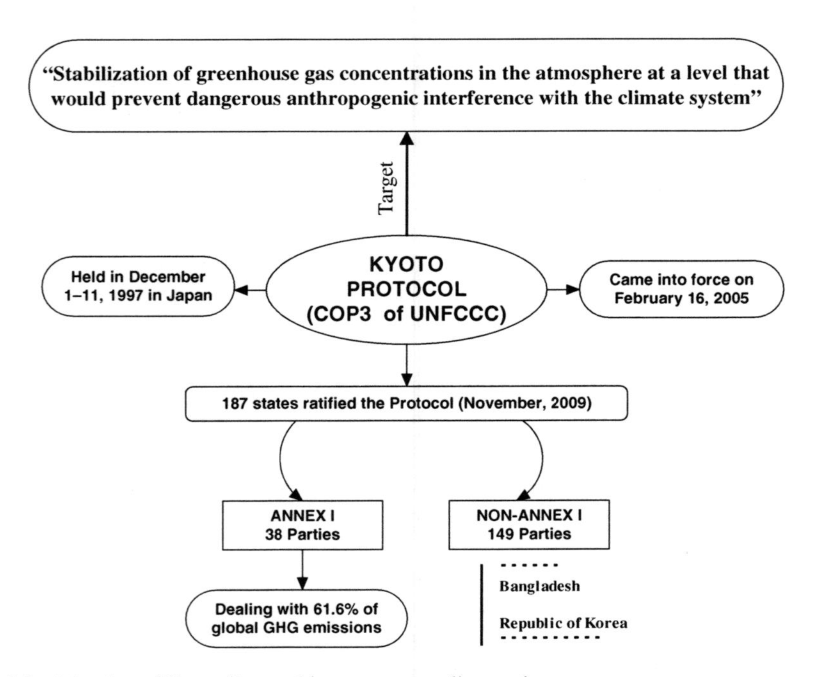


**Fig. 1.2** Adoption of Kyoto Protocol in response to climate change

Protocol Parties. The total percentage of the Annex I Parties' emissions is 61.6% (UNFCCC 2006a). Bangladesh and the Republic of Korea remain in the Non-Annex I list of the Parties.

Several provisions in the UNFCCC and the Kyoto Protocol allow nations to achieve GHG emission reductions or enhancement of sinks cooperatively. As a general matter, the UNFCCC commits the most highly developed nations to give developing countries financial and technical assistance to implement the Convention and to deal with the effects of climate change. Article 4 of the UNFCCC obligates developed nations to assist developing nations through funding for emission reductions, funding for adaptation to adverse effects, and transfer of environmentally sound technology (Rosenbaum et al. 2004). Articles 4, 6, 12, and 17 of the Kyoto Protocol contemplate flexible mechanisms of compliance. Article 4 deals with the possibility that a group of Annex I Parties or a regional economic integration organization could jointly fulfill their reduction commitment. Article 6 allows Annex I Parties to transfer 'emission reduction units' generated through JI (Joint Implementation) projects and allows Parties to authorize 'legal entities' to participate in these transfers. Eligible JI projects include all 'Land Use, Land-Use Change and Forestry' (LULUCF) activities allowed under Article 3.

### *1.2.3 CDM Helps Attain GHG Reduction Commitment (Annex I) and Derive Sustainable Development in the Host (Non-Annex I) Countries*

Article 12 of the Kyoto Protocol introduces the Clean Development Mechanism (CDM), originally a part of Activities Implemented Jointly (AIJ). The CDM is an instrument under the authority of the COP and supervised by an executive board. CDM projects typically involve Annex I countries as investors and Non-Annex I countries as hosts, essentially joint ventures between developed and developing countries (Fig. 1.3).

Reductions resulting from these projects, beginning in the year 2000, count toward satisfying an Annex I country's obligations to reduce aggregate emissions during the years 2008–2012 (first commitment period). An 'operational entity' accredited by the COP must validate the project before implementation and verify the project's emission reductions before the executive board can issue credits for the emission reductions achieved.

According to the modalities of the CDM adopted at COP7 in Marrakesh, Morocco, in November 2001 (Marrakesh Accords), it can allow projects both in LULUCF and in energy sectors. In the first commitment period, CDM restricts the LULUCF projects only to A/R which comply with the Subsidiary Body for Scientific and Technological Advice (SBSTA) recommendations adopted in COP9 (Decision 19/CP.9). The recommendation of SBSTA includes the details of definitions, non-permanence, leakage, additionality, uncertainties, and socio-economic and environmental impacts, including impacts on biodiversity and natural

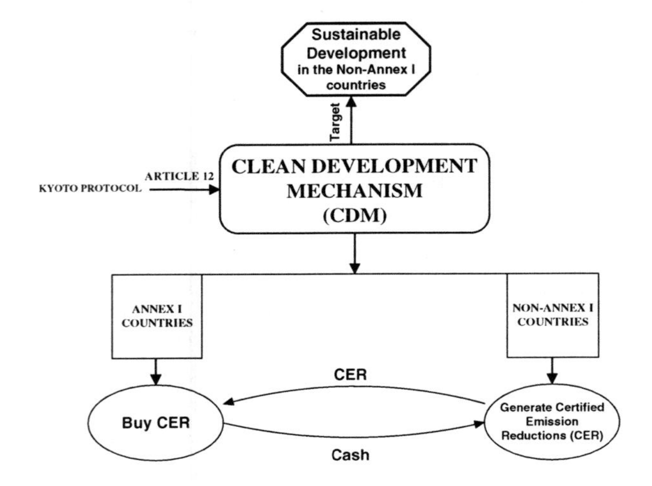


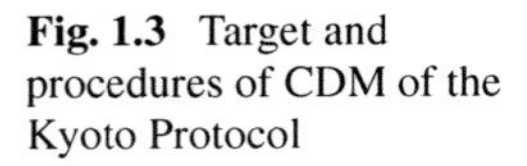
**Fig. 1.3** Target and procedures of CDM of the Kyoto Protocol

ecosystems (Rosenbaum et al. 2004). Silveira (2005) discusses the role of CDM in respect to sustainable development, formation of carbon markets, and promotion of bioenergy options. His study concludes that bioenergy projects are attractive and CDM provides a complementary bridge for international cooperation toward sustainable development. Ravindranath et al. (2006) and Reddy & Balachandra (2006) also conclude that a wood fuel stove project with the improvement of the traditional stoves can well be put on the international 'carbon market' at competitive cost for GHG emission reduction. Forest management and conservation as well as carbon sequestration in agriculture are not allowed in the first commitment period. Furthermore, the credit that a Party can claim from LULUCF projects under the CDM is 1% of the Party's base year (1990) emissions times five (Rosenbaum et al. 2004). The COP agreement means that over the 5 years, one-fifth of the reduction can come from CDM-LULUCF projects (Rosenbaum et al. 2004). Negotiations are being continued as to how to treat CDM-LULUCF projects after 2012.

However, CDM projects are expected to derive sustainable development in the Non-Annex I countries (Fig. 1.4). The development must be in the social, environmental, and economic arena of a country. The possible carbon sequestration, carbon combustion efficiency, and carbon substitution projects are expected to derive lots of impacts on many sectors of a host country (Fig. 1.5).

### *1.2.4 Carbon Sequestration in the Forests Can Cost-Effectively Reduce the Atmospheric GHGs*

Terrestrial ecosystems clearly influence the concentration of GHGs to the atmosphere (Lal 2004) as they work as both sinks and sources. Greenhouse gases are constantly entering and leaving the atmosphere. Actively growing trees and

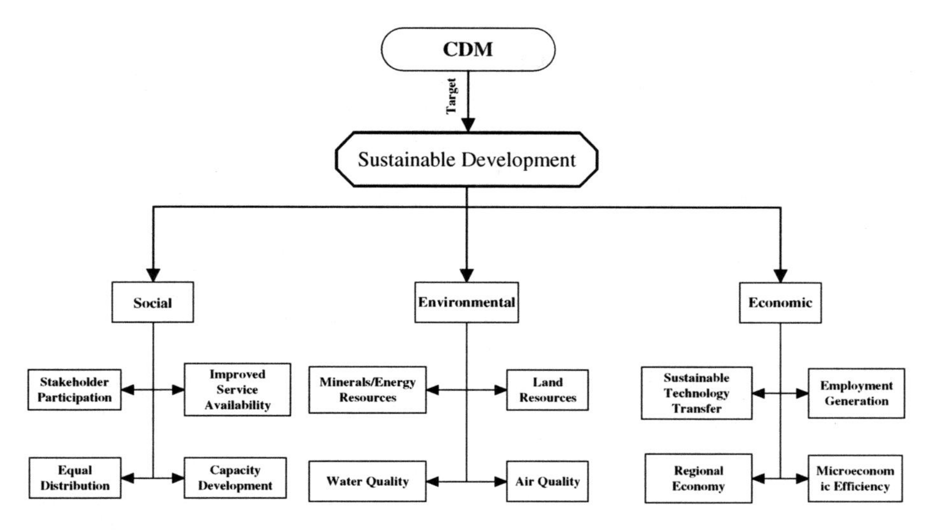


**Fig. 1.4** Sustainable development criteria of the CDM of the Kyoto Protocol

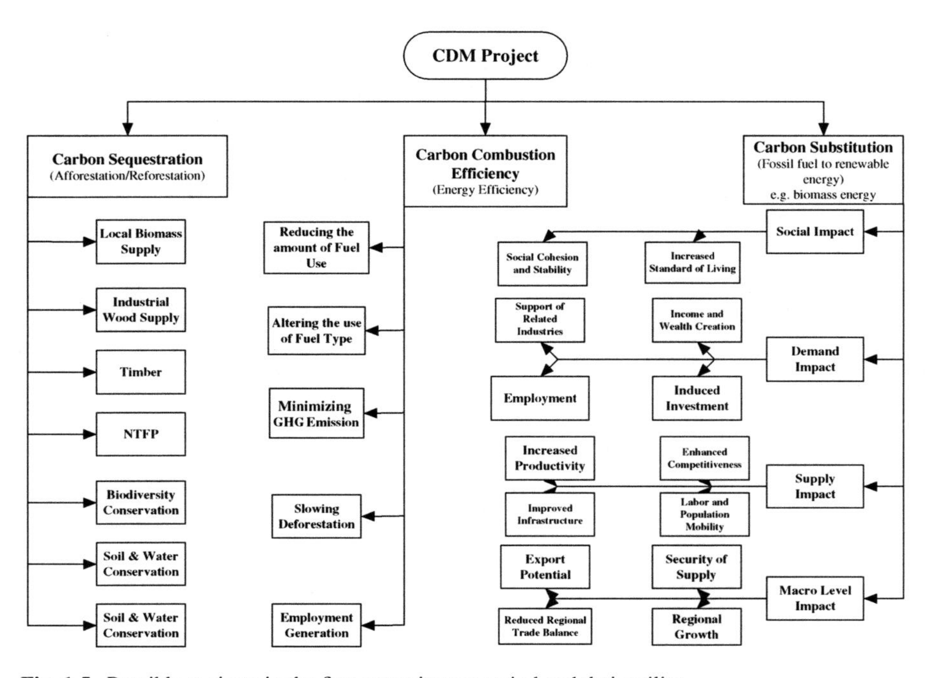


**Fig. 1.5** Possible projects in the first commitment period and their utility

other plants uptake $CO_2$ from the atmosphere, combine it with water through photosynthesis, and create sugars and more stable carbohydrates (Rosenbaum et al. 2004). Through this process, trees capture and store atmospheric $CO_2$ in vegetation, soils, and biomass products. Carbohydrates become the building blocks and energy supply for most of the life on Earth. Eventually, when plants and animals die, $CO_2$ returns to the atmosphere. When wood products or other organic materials burn or decompose, they also release $CO_2$ (Fig. 1.6).

The terrestrial ecosystem plays an important role in global carbon sequestration. It has been estimated that 1,146 GtC is stored within the 4.17 b ha of tropical, temperate, and boreal forest areas, about one-third of which is stored in forest vegetation (IPCC 2000). Aboveground biomass in the tropical forests and belowground biomass of the savannas in the tropics have the greatest carbon storage at 212 and 264 GtC, respectively (IPCC 2000). Soil carbon represents the largest carbon pool of terrestrial ecosystems and has been estimated to have one of the largest potentials to sequester carbon worldwide (García-Oliva & Masera 2004). Soil stores about 80% of the total carbon sequestered by the terrestrial ecosystem, ranging from 50% in the tropical forests to 95% in the tundra (IPCC 2000). The anthropogenic causes hamper this storage of carbon in the forests.

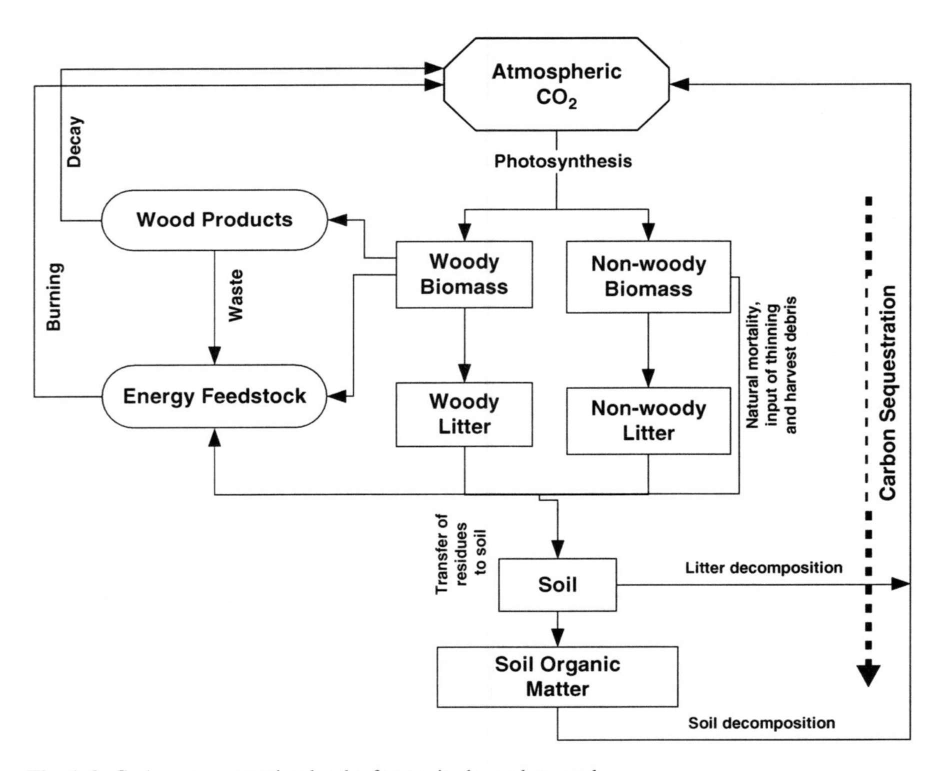


**Fig. 1.6** Carbon sequestration by the forests in the carbon cycle

The practice of sustainable forest management can enhance the sinking characteristics of forests (Rosenbaum et al. 2004). Establishment, enhancement, and protection of forest ecosystems can affect the GHG concentration in the atmosphere. A/R of non-forested or degraded forestlands can increase; and prevention of deforestation can maintain the amount of carbon held in forests (Rosenbaum et al. 2004). The relative low cost of A/R, compared with non-forest offset options, may make them economically attractive (Cannell 2003; Niles et al. 2002; Ravindranath & Somashekhar 1995). Selective cutting schemes, lengthened rotations, reduced-impact logging, and species choice may achieve a higher average level of sequestered carbon (Rosenbaum et al. 2004). Simply postponing or eliminating harvesting can sometimes be a short- to medium-term means to keep carbon sequestered (Schulze et al. 2000).

Using wood in buildings and other long-lived objects effectively sequesters carbon for the life of the objects. Substituting essentially carbon-neutral wood for energy-intensive materials such as brick, aluminum, or steel may significantly reduce the use of fossil fuels, which of course release $CO_2$ when burned (Rosenbaum et al. 2004).

Sustainable production of wood fuel from forests can displace fossil fuels. Although burning of biomass fuels releases $CO_2$, the regrowth of a sustainably managed forest offsets that release. Thus, forest fuels can supply energy virtually without net contribution to GHG levels (Rosenbaum et al. 2004).

### *1.2.5 Biomass Combustion Efficiency Can Help Minimize GHG Emission and Slow Deforestation*

Biomass fuels are the basic energy source in developing countries. Cooking energy constitutes a major fraction of energy consumption in rural areas of developing countries, which is largely met with biofuels, such as fuelwood, charcoal, agri-residue, and dung cake (Rubab & Kandpal 1996).

It has been estimated that biomass energy accounts for about 15% of the world's energy consumption and about 38% of the primary energy consumption in developing countries (Sims 2003). It has also been estimated that biomass often accounts for more than 90% of the total rural energy supplies in developing countries (Bhattacharya & Salam 2002). Among the fuels used for domestic energy in Bangladesh, tree and bamboo provide 48%, agricultural residues 36%, dung 13%, and peat 3% (GOB 1993). Developed countries use more non-forest fuel energy than developing countries (Ravindranath et al. 2006; Sims 2003). Among biofuel use, wood fuel is a dominant domestic fuel in both rural and urban areas of the developing countries (Arnold et al. 2006; Jashimuddin et al. 2006; Miah et al. 2003; Ouedraogo 2006).

Traditional biomass-fired stoves have been identified as inefficient due to its incomplete combustion system (Qiu et al. 1996; Rubab & Kandpal 1996). Bhattacharya & Salam (2002) report that it causes significant GHG emissions due to formation of products of incomplete combustion and also poses health hazards. Qiu et al. (1996) also report that due to the unsustainable biomass production, it causes ecological and environmental problems, such as deforestation and land degradation.

Although GHG emission and health hazards from the traditional cooking stoves were not of much concern in the last decades, the growing scarcity of biofuels, which was partly due to their inefficient utilization in traditional cooking stoves, had been the primary motivating factor for the development and dissemination of improved biofuel cooking stoves (Hyman 1994; Rubab & Kandpal 1996). Bhattacharya & Salam (2002) report that new stove designs can improve the efficiency of biomass use for cooking by a factor of 2–3; and the amount of biomass saved through increase in biomass use efficiency can reduce GHG emission further through substitution of fossil fuels.

Bhattacharya et al. (1999) studied the 'potential of biomass fuel conservation in selected Asian countries' including China, India, Nepal, Pakistan, Philippines, Sri Lanka, and Vietnam. They estimated that through the efficiency improvements in these countries, total biomass fuels saved were 326 m t annually. They also estimated that if all the traditional stoves can be converted to improved ones, the total

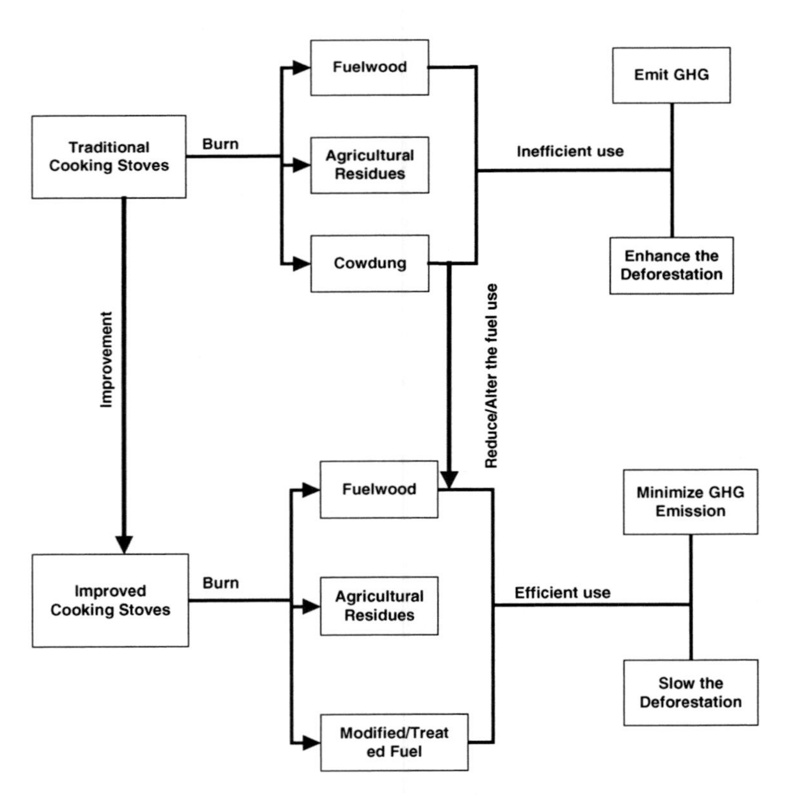


**Fig. 1.7** Biomass combustion efficiency in the CDM of the Kyoto Protocol

saved biomass can be 296 m t annually. In total, the improved ones can save around 152 m t of fuelwood in the domestic sector which is about 43% of the total fuelwood use in the domestic sector. In Bangladesh, a severe dependence on biofuels by a large number of people has a substantial potential for diffusion of improved cooking stoves, which can save the fuelwood thereby slowing the deforestation and minimizing the GHG emission (Fig. 1.7).

## 1.3 Scopes of the Book

Based on the theoretical framework, this book aimed at resolving the following research questions (Fig. 1.8):

1. What potential mitigation options exist in the forestry sector of Bangladesh?
2. What are the biomass energy promotion options that exist in the forestry sector of Bangladesh?
3. How can carbon be sequestered in the CDM-A/R projects in Bangladesh?
4. What are the implications of the traditional cooking stove with the GHG emission in Bangladesh?
5. What national policy issues does Bangladesh need to comply with the forestry activities to the CDM?

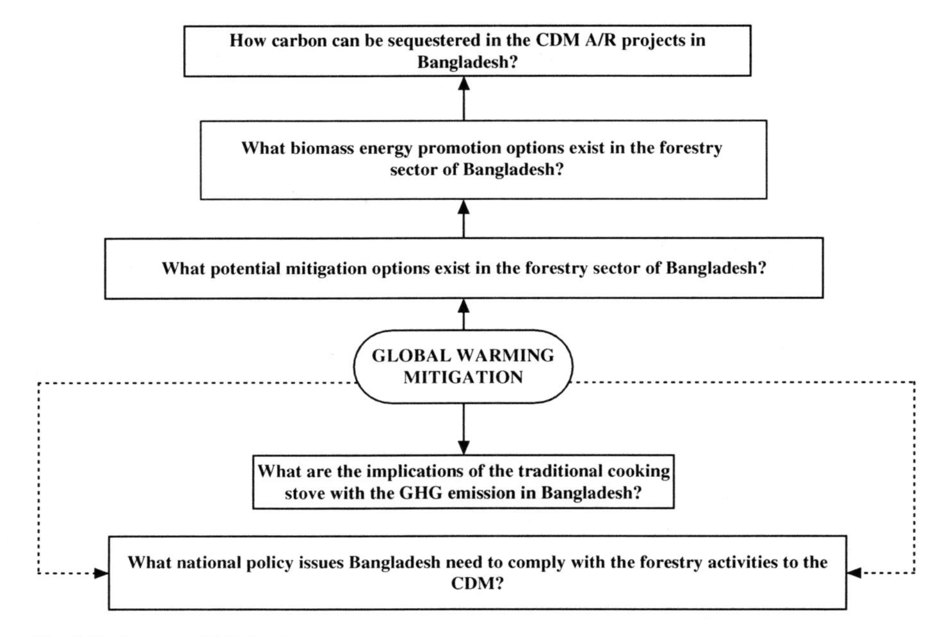


**Fig. 1.8** Scopes of this book

## 1.4 Objectives of This Publishing Attempt

The general purpose of the study is to find out the means and ways to contribute efficiently to reducing global warming through the forestry options in Bangladesh with the implications of CDM.

The specific objectives are

- To find out the climate change mitigation options of the forestry sector
- To find out the options of biomass energy promotion
- To identify the carbon sequestration induced by reforestation in Bangladesh
- To figure out the traditional use of biomass in the cooking stove and GHG emission
- To figure out the reforestation success of the Republic of Korea and offering suggestions to Bangladesh
- To discuss the legal framework of the CDM of the Kyoto Protocol for adjusting the existing forest policy of the countries to achieve the carbon credits

## 1.5 General Methodology

The book attempts to address the three important issues, i.e., climate change mitigation options of the forestry sector of Bangladesh (Chap. 3), biomass energy promotion in Bangladesh (Chap. 4), and carbon sequestration potential in the expected A/R projects in Bangladesh (Chap. 5). All of the issues have focused on the CDM of the Kyoto Protocol. Chapter 2 discusses on the general overview of the CDM. Chapter 3 deals with the critical analysis to find out the global warming mitigation options in the forestry sector of Bangladesh. The raw data collected from the Forestry Directorate and other sources in Bangladesh have been analyzed to find out the suitable forestry options which can comply with the CDM modalities. As the reforestation success of the Republic of Korea has been a global reforestation model (Brown 2006), the book has figured out this fact which can be a lesson for Bangladesh. Chapter 4 deals with the critical analysis of the data collected from the various sources to show how Bangladesh can promote biomass as a global warming response. As traditional cooking stoves in Bangladesh have been recognized as inefficient in biomass burning process, this chapter finds out innovative traditional cooking stoves which can be promoted through the CDM activities. In Chap. 5, the study analyzed the data to show the carbon sequestration potential of Bangladesh in general and of Chittagong region in particular. To show the specific scenario of Chittagong region, the study collected the data for 13 tree species, with ages ranging from 6 to 23 years. Biomass was accounted using the accepted allometric regression equation. Litter and soil carbon were analyzed using the appropriate procedure described in Chap. 5. Mean Annual Increment (MAI) was also accounted considering the carbon loss from the stands. In Chap. 6, the study concludes all the findings and intends to recommend several ways and means to effectively participate in the global warming mitigation options by the forestry options in Bangladesh. All the quantitative data were processed and analyzed using SPSS 13.0 for Windows.

# Chapter 2
# General Overview of the Clean Development Mechanism (CDM)

**Abstract** The CDM is an important flexible mechanism of the Kyoto Protocol. Understanding general overview of the CDM is critical to dealing with CDM forestry. The CDM regulators, CDM project cycle, sustainable development issues of the CDM are the important common issues of the CDM. This chapter discusses these issues importantly. Present global status of CDM projects also have been described here.

## 2.1 Introduction

Article 12 of the Kyoto Protocol introduces the Clean Development Mechanism (CDM) as one of the three 'flexible' mechanisms. The CDM is an instrument under the authority of the COP and supervised by an Executive Board (EB). CDM projects typically involve Annex I countries as investors and Non-Annex I countries as hosts, essentially joint ventures between developed and developing countries. The formulations of this mechanism have twofold objectives, i.e., reducing greenhouse gases and contributing to the sustainable development in the host countries (Olsen and Fenhann 2008). It also assists Annex I countries in achieving their emission reduction targets in a cost-effective manner. This chapter deals with the general overview of the CDM.

## 2.2 The CDM Regulators

The COP/MOP, EB, Designated Operational Entity (DOE), Host Party, and Donor Party are the important regulators for the CDM. The responsibility of the COP/MOP is to accredit the standards for and designation of the DOE. It reviews the regional/subregional distribution of the CDM project activities. It also oversees the rules and procedures of EB. The EB is responsible for approving the methodologies for baselines and monitoring plans and project boundaries. It maintains the registry of CDM. It also accredits the DOEs. The COP/MOP designates and the EB accredits

Md.D. Miah et al., *Forests to Climate Change Mitigation*, Environmental Science and Engineering, DOI 10.1007/978-3-642-13253-7_2, © Springer-Verlag Berlin Heidelberg 2011 

the DOE. They are responsible for validating project design document (PDD). They also verify and certify the ERs (Emission Reductions). Host Party is certainly a country which is in the Non-Annex I list and ratified the Kyoto Protocol. They are responsible for creating a Designated National Authority (DNA) for the CDM. The Donor Party must be in the list of Annex I parties that ratified the Kyoto Protocol. They must have the targets of emission reductions as Article 3 of the Kyoto Protocol. They should also have the national system for estimating sources and sinks of GHG as Article 5 of the Kyoto Protocol.

## 2.3 The CDM Project Cycle

To attain the CERs (Certified Emission Reductions) in the host countries, the CDM project must have to maintain the specific procedures agreed in the Marrakech Accords. The Project Participant (PP), the Host Party, must have to create an idea of the CDM Project which is called Project Design Document (PDD). It requires the validation by the DOE, preferably by the Designated National Authority (DNA). It is then sent to the EB for registration. After registration, the project can run in the host country. According to the approved plan, monitoring is also a required step for measuring ERs. The DNA then verifies and certifies the ERs which are finally issued by the EB. The PP receives the certificate of CERs at this stage which can be sold out in the carbon market (Fig. 2.1).

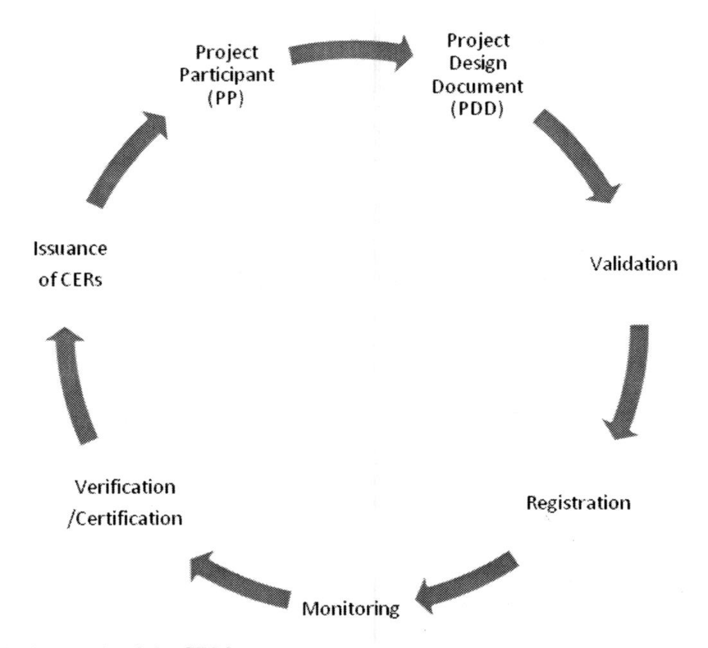


**Fig. 2.1** Project cycle of the CDM

However, the PDD needs to fulfill all of the requirements, i.e., definition of project boundaries, assessment of country context, assessment of additionality, definition of crediting lifetime, projection of baseline scenario, monitoring of project, calculation of ER, correction of leakage, and uncertainty in the ER.

## 2.4 Small-Scale CDM

As the carbon benefits from the small-scale CDM (SSC) project are comparatively shorter than the standard CDM project, small-scale CDM project is attractive in the developing countries like Bangladesh. In the COP8 in 2002, India, a report on streamlined modalities for small-scale projects was produced by the expert group. The differences between the small-scale and standard CDM project lie in baseline methodologies assigned to the project types. Time spent on the baseline formulation in the small-scale CDM is dramatically reduced. If suitable baseline is unavailable, then a proposed approach can be submitted to the EB for approval. Correction for leakage is not necessary in the small-scale CDM project. Additionality is expressed mainly based on the barriers they face. The calculation of the ER is very simplified and the transaction cost becomes so lowered. Environmental Impact Assessment (EIA) is not necessary in this smaller CDM. There are three types of SSC projects approved by the EB. The type I project activities, such as renewable energy project activities, shall have a maximum output capacity of 15 MW or an equivalent (IGES 2009). The type II project activities are relevant to the improvements in energy efficiency, which reduce energy consumption on both the demand and supply side. But the maximum output is limited to 60 GWh[1] (or an appropriate equivalent) per year. The type III projects are defined as projects other than that of type I and type II. The ER should be less than or equal to 60 kt[2] $CO_2$ equivalent annually. In type I, projects can be electricity generation by user/household, mechanical energy for the user/enterprise, thermal energy for the user, and electricity generation for a system. In type II, supply-side energy efficiency improvement in transmission and distribution, supply-side energy efficiency improvement in generation, demand-side energy efficiency programs for specific technologies, energy efficiency and fuel switching measures for industrial activities, and energy efficiency and fuel switching measures for building can be important. Agriculture, forestry, switching fossil fuels, emission reductions in the transport sector, and methane recovery are important fields for type III projects.

[1] 60 GWh is equivalent to 4,000 h of operation of a 15 MW plant or $60 \times 3.6$ TJ = 216 TJ. TJ = terajoules.

[2] Kiloton.

## 2.5 Sustainable Development Issues in the CDM Projects

The major target of the CDM project is to generate sustainable development (SD) in the developing countries. Misana and Karlsson (2001) and Olmos (2001) found a strong relationship between the CDM project and SD in the developing countries. Begg et al. (2000) proved that small-scale CDM projects in the domestic sector of LDCs (Least Developed Countries) can have significant SD benefits, such as freeing up time and energy for other activities, e.g., economic, cultural, and educational. It also can save money and improve the living standards. However, indicators can be used to prioritize projects on the basis of SD criteria. Olsen and Fenhann (2008) analyzed the sustainability issues of the CDM projects using 744 PDD submitted for validation by May 3, 2006. Analyzing all the PDDs, they found the most common five benefits of CDM projects in the developing countries. These are employment generation, economic growth, a better quality of air, access to energy, and welfare improvement. The generation of SD benefits varies from project to project. Their analysis confirmed that few SD benefits are generated from HFC and $N_2O$ projects. CDM projects on energy efficiency in the industrial sector also have few SD benefits, especially a higher contribution to the improved air quality. Olsen and Fenhann (2008) also found that renewable energy generation, especially biomass energy, did not find so much SD benefits. But wind and hydro projects had comparatively higher SD benefits (employment, welfare, growth, and access to energy) than that of the biomass energy. They found that $CH_4$ reduction projects contribute slightly higher number of SD benefits than the renewable energy categories. They found that cement projects contributed with so many SD benefits, with 82% of all the projects contributing to better air quality and conservation. They also found that small-scale projects delivered comparatively higher SD benefits than that of the large-scale projects.

## 2.6 Present Status of CDM Projects

Up to February 1, 2010, the total number of registered CDM projects is 2,029 (IGES 2010). Among them small-scale projects are 903 (44%) and the large-scale projects are 1,126 (56%) in number (Table 2.1). However, the CDM projects deal with the 1,785,802,000 t-$CO_2$e total emission reductions by 2012. Among them the small-scale projects deal with 135,013,000 t-$CO_2$e and the large-scale projects 1,650,789,000 t-$CO_2$e emission reductions by 2012. Most of the CDM projects, 27%, are on hydropower generation, reducing 209,171,000 t-$CO_2$e total emission reductions by 2012. Biogas, wind power, and biomass-based CDM projects represent 14.39, 13.95, and 12.47%, respectively, in the total projects registered. A/R CDM projects only represent 0.64% (Table 2.2). The distribution of CDM projects shows that 73.24% projects are distributed in the Asian countries followed by 22.38% in the Latin American countries (Table 2.3). China and India represent the greatest share of the CDM projects within Asia, 49.26 and 32.37%, respectively. In the global perspective, they also have the greatest share, 36.08 and 23.37%, respectively. Bangladesh only has 0.13% CDM projects within Asia, with the expected 1,191,000 t-$CO_2$e total emission reductions by 2012.

**Table 2.1** Small-scale and large-scale registered CDM projects up to February 1, 2010, in the developing countries

| Size of the CDM project | Annual emission reductions (t-$CO_2$/year) (UNFCCC) | Number of CDM projects | Total emission reductions by 2012 (1,000 t-$CO_2$e) |
|---|---|---|---|
| Small scale | 0–50,000 | 821 | 96,354 |
| | 50,000–100,000 | 67 | 19,089 |
| | 100,000–150,000 | 5 | 3,274 |
| | 150,000–200,000 | 2 | 3,596 |
| | 200,000–250,000 | 5 | 6,967 |
| | 250,000–300,000 | 2 | 3,518 |
| | 350,000–400,000 | 1 | 2,214 |
| **Small-scale subtotal** | | **903** | **135, 013** |
| Large scale | 0–50,000 | 213 | 41, 564 |
| | 50,000–100,000 | 324 | 121, 192 |
| | 100,000–150,000 | 227 | 127, 748 |
| | 150,000–200,000 | 97 | 80, 899 |
| | 200,000–250,000 | 44 | 44, 891 |
| | 250,000–300,000 | 41 | 55, 838 |
| | 300,000–350,000 | 29 | 41, 813 |
| | 350,000–400,000 | 20 | 32, 245 |
| | 400,000–450,000 | 12 | 24, 122 |
| | 450,000–500,000 | 13 | 28, 458 |
| | 500,000–550,000 | 9 | 19, 889 |
| | 550,000–600,000 | 10 | 29, 874 |
| | 600,000–650,000 | 6 | 17, 413 |
| | 650,000–700,000 | 10 | 41, 986 |
| | 700,000–750,000 | 1 | 3, 125 |
| | 750,000–800,000 | 7 | 28, 141 |
| | 800,000–850,000 | 4 | 16, 732 |
| | 850,000–900,000 | 3 | 10, 335 |
| | 900,000–950,000 | 7 | 26, 822 |
| | 950,000–1,000,000 | 2 | 10, 909 |
| | >1,000,000 | 47 | 846, 793 |
| **Large-scale subtotal** | | **1,126** | **1,650,789** |
| **Total** | | **2,029** | **1,785,802** |

Source: IGES (2010)

**Table 2.2** Types of CDM projects registered up to February 01, 2010, and emission reduction by 2012

| CDM project type | Number of CDM projects | Percentage of the CDM project among the total projects registered | Emission reductions by 2012 (1,000 t-$CO_2$e) |
|---|---|---|---|
| HFC reduction/avoidance | 20 | 0.99 | 484, 567 |
| $N_2O$ reduction | 60 | 2.96 | 252, 268 |
| PFC reduction | 3 | 0.15 | 1, 817 |
| $SF_6$ replacement | 3 | 0.15 | 2, 052 |
| Cement | 29 | 1.43 | 28, 626 |
| Other renewable energies | 31 | 1.53 | 13, 014 |
| Biogas | 292 | 14.39 | 64, 122 |
| Biomass | 253 | 12.47 | 89, 801 |
| Methane recovery and utilization | 163 | 8.03 | 214, 855 |
| Methane avoidance | 43 | 2.12 | 8, 166 |
| Transportation | 2 | 0.10 | 1, 963 |
| Energy efficiency | 75 | 3.70 | 18, 222 |
| Afforestation and reforestation | 13 | 0.64 | 2, 152 |
| Hydropower | 552 | 27.21 | 209, 171 |
| Fuel switch | 54 | 2.66 | 110, 751 |
| Waste gas/heat utilization | 152 | 7.49 | 156, 335 |
| Wind power | 283 | 13.95 | 126, 562 |
| Leak reduction | 1 | 0.05 | 1, 357 |
| Total | 2, 029 | 100.00 | 1, 785, 802 |

Source: IGES (2010)

**Table 2.3** Countrywise registered CDM projects up to February 01, 2010

| Region | Country | Number of CDM projects | Regional percentage of the CDM projects | Global percentage of the CDM projects | Emission reductions by 2012 (1,000 t-$CO_2$e) |
|---|---|---|---|---|---|
| Asia | India | 481 | 32.37 | 23.71 | 255, 729 |
| | Indonesia | 43 | 2.89 | 2.12 | 21, 564 |
| | Cambodia | 4 | 0.27 | 0.20 | 604 |
| | Singapore | 1 | 0.07 | 0.05 | 74 |
| | Sri Lanka | 6 | 0.40 | 0.30 | 1, 343 |
| | Thailand | 30 | 2.02 | 1.48 | 10, 416 |
| | Nepal | 2 | 0.13 | 0.10 | 744 |
| | Pakistan | 4 | 0.27 | 0.20 | 7, 121 |
| | Papua New Guinea | 1 | 0.07 | 0.05 | 1, 834 |
| | Bangladesh | 2 | 0.13 | 0.10 | 1, 191 |
| | Fiji | 1 | 0.07 | 0.05 | 199 |
| | The Philippines | 40 | 2.69 | 1.97 | 6, 159 |
| | Bhutan | 1 | 0.07 | 0.05 | 4 |

**Table 2.3** (continued)

| Region | Country | Number of CDM projects | Regional percentage of the CDM projects | Global percentage of the CDM projects | Emission reductions by 2012 (1,000 t-$CO_2$e) |
|---|---|---|---|---|---|
| | Vietnam | 20 | 1.35 | 0.99 | 9,483 |
| | Malaysia | 79 | 5.32 | 3.89 | 22,489 |
| | Mongolia | 3 | 0.20 | 0.15 | 386 |
| | Lao PDR | 1 | 0.07 | 0.05 | 19 |
| | South Korea | 35 | 2.36 | 1.72 | 95,470 |
| | China | 732 | 49.26 | 36.08 | 964,814 |
| **Asia subtotal** | | **1486** | 100.00 | 73.24 | **1,399,643** |
| Africa/Middle and Near East | Arab United Emirates | 4 | 6.25 | 0.20 | 897 |
| | Israel | 16 | 25.00 | 0.79 | 8,811 |
| | Iran | 1 | 1.56 | 0.05 | 1,632 |
| | Uganda | 2 | 3.13 | 0.10 | 275 |
| | Egypt | 4 | 6.25 | 0.20 | 10,486 |
| | Ethiopia | 1 | 1.56 | 0.05 | 108 |
| | Qatar | 1 | 1.56 | 0.05 | 13,984 |
| | Kenya | 1 | 1.56 | 0.05 | 551 |
| | Republic of Cote d'Ivoire | 1 | 1.56 | 0.05 | 178 |
| | Zambia | 1 | 1.56 | 0.05 | 357 |
| | Syria | 2 | 3.13 | 0.10 | 501 |
| | Tanzania | 1 | 1.56 | 0.05 | 986 |
| | Tunisia | 2 | 3.13 | 0.10 | 3,741 |
| | Nigeria | 3 | 4.69 | 0.15 | 18,712 |
| | Morocco | 5 | 7.81 | 0.25 | 1,718 |
| | Jordan | 2 | 3.13 | 0.10 | 2,300 |
| | South Africa | 17 | 26.56 | 0.84 | 15,926 |
| **Africa/Middle and Near East subtotal** | | **64** | 100.00 | 3.15 | **81,163** |
| | Albania | 1 | 4.00 | 0.05 | 44 |
| | Armenia | 5 | 20.00 | 0.25 | 1,169 |
| | Uzbekistan | 7 | 28.00 | 0.34 | 4,076 |
| Others | Cyprus | 5 | 20.00 | 0.25 | 507 |
| | Georgia | 2 | 8.00 | 0.10 | 1,642 |
| | Macedonia | 1 | 4.00 | 0.05 | 168 |
| | Moldova | 4 | 16.00 | 0.20 | 1,254 |
| **Others subtotal** | | **25** | 100.00 | 1.23 | **8,861** |
| Latin America | Argentina | 16 | 3.52 | 0.79 | 27,359 |
| | Uruguay | 3 | 0.66 | 0.15 | 1,414 |
| | Ecuador | 14 | 3.08 | 0.69 | 3,835 |
| | El Salvador | 5 | 1.10 | 0.25 | 3,426 |

**Table 2.3** (continued)

| Region | Country | Number of CDM projects | Regional percentage of the CDM projects | Global percentage of the CDM projects | Emission reductions by 2012 (1,000 t-$CO_2$e) |
|---|---|---|---|---|---|
| | Guyana | 1 | 0.22 | 0.05 | 222 |
| | Cuba | 2 | 0.44 | 0.10 | 2,463 |
| | Guatemala | 11 | 2.42 | 0.54 | 4,629 |
| | Costa Rica | 6 | 1.32 | 0.30 | 2,292 |
| | Colombia | 20 | 4.41 | 0.99 | 13,330 |
| | Jamaica | 1 | 0.22 | 0.05 | 455 |
| | Chile | 36 | 7.93 | 1.77 | 29,567 |
| | Dominican Republic | 1 | 0.22 | 0.05 | 299 |
| | Nicaragua | 4 | 0.88 | 0.20 | 4,666 |
| | Panama | 6 | 1.32 | 0.30 | 1,407 |
| | Paraguay | 1 | 0.22 | 0.05 | 24 |
| | Brazil | 168 | 37.00 | 8.28 | 136,379 |
| | Peru | 21 | 4.63 | 1.03 | 8,983 |
| | Bolivia | 3 | 0.66 | 0.15 | 2,503 |
| | Honduras | 15 | 3.30 | 0.74 | 2,162 |
| | Mexico | 120 | 26.43 | 5.91 | 50,720 |
| **Latin America subtotal** | | **454** | 100.00 | 22.38 | **296,134** |
| **Total** | | **2029** | | 100.00 | **1,785,802** |

Source: IGES (2010)

# Chapter 3
# CDM Forests in Bangladesh and Learning from the Reforestation Success of the Republic of Korea

**Abstract** The CDM is an important economic tool to mitigate global climate change and support sustainable development in the non-Annex I countries. Bangladesh, a non-Annex I country, can be a potential host country of the forests-CDM projects, because of the potentialities of the carbon sequestration in the forests. This chapter finds that afforestation and reforestation (A/R) may be one of the greatest choices in conserving the existing carbon sink because it offers the opportunities of carbon credits that are subject to the end use of the forest products. The chapter focuses the legal frameworks of the A/R CDM activities and critical issues which should be resolved by the policy interventions in Bangladesh. The study also figures out the reforestation success of the Republic of Korea and suggests to Bangladesh what can be learnt from the Republic of Korea. The outcome of this study will be of great importance to the policy makers in the field of forest restoration in Bangladesh and global warming mitigation.

## 3.1 Introduction

Global climate change is one of the critical issues in the world (Houghton 2005). Forest ecosystems, which play important roles in climate change, can assimilate $CO_2$ via photosynthesis and also store carbon in their biomass and in soil (IPCC 2000). Tropical forests which make up 80% of the total forests on the Earth are considered to have the biggest long-term potential to sequester atmospheric carbon by protecting forested lands, slowing down deforestation rate, and encouraging A/R (Brown et al. 1996). It is estimated that around 1.8 GtC $yr^{-1}$ moves from the world's forests to the atmosphere due to deforestation, harvesting, and forest degradation, and 20% of that carbon source is from tropical deforestation (IPCC 2000).

Forestry and land-use change activities contribute to reducing GHG emissions by A/R, avoiding deforestation, and improving forest management (Pearson et al. 2005). Under the agreement made at the COP9, industrialized (Annex I) countries can partially meet their emission reduction commitments under the Kyoto Protocol by financing A/R in developing countries through the CDM (UNFCCC 2006b).

Md.D. Miah et al., *Forests to Climate Change Mitigation*, Environmental Science and Engineering, DOI 10.1007/978-3-642-13253-7_3, © Springer-Verlag Berlin Heidelberg 2011

Bangladesh is assumed to play an important role in mitigating global warming with a huge pool of existing plantations and natural forests. In addition, Bangladesh has wide areas of degraded forestlands and other wastelands, which can be reforested. However, the severe poverty and the lack of appropriate technology in Bangladesh show the 'additionality' for A/R activities in the CDM. Therefore, Bangladesh can effectively participate in carbon trading. However, it is necessary to develop national policies in mitigating global warming to bring about large-scale changes in land-use and forestry practices and to address some of the technical and policy issues that have proven to be particularly problematic from carbon-accounting and project-level perspectives (Kennett 2002). But the country lacks appropriate policies and research on different forestry options for mitigating global warming. The Republic of Korea has placed a significant example in the world in the arena of restoration of the degraded forestlands (Brown 2006). Within the last three decades, the country has changed the national forest land use very positively within a comparatively shorter time which can be a reforestation model or reforestation learning to the developing countries with mass degradation in the forestlands.

It is necessary to understand the potential of different forestry options for mitigating global warming, unsettled issues to be met by the forestry sector, to participate in the CDM forestry activities. The study aims to figure out different forestry mitigation options and to discuss the important unresolved issues in carbon sequestration. This study sketches out the policy issues in the future of the forestry sector in mitigating global warming. It also figures out the reforestation success in the Republic of Korea which can be a lesson to Bangladesh. The findings of this study will be of great use to the policy makers in the arena of global warming mitigation in Bangladesh and in other countries as well.

## 3.2 The Legal Issues of Mitigating Global Warming

The Kyoto Protocol was adopted in December 1997 at COP3 in Kyoto, Japan, which came into force on February 16, 2005. In that Protocol, the developed countries and the countries in transition to a market economy were committed to achieve emissions reduction targets. These countries, known under the UNFCCC as Annex I Parties, agreed to reduce their overall emissions of six GHGs by an average of 5.2% below 1990 levels between 2008 and 2012 (the first commitment period), with specific targets varying from country to country. The Protocol also establishes three flexible mechanisms to assist Annex I Parties in meeting their national targets cost-effectively, i.e., emission trading system; joint implementation (JI) of emissions reduction projects between Annex I Parties; and the CDM, which allows for emissions reduction projects to be implemented in Non-Annex I Parties (developing countries). The CDM was established by Article 12 of the Kyoto Protocol to create 'certified emission reductions' (CER), generated by projects in developing countries without emissions limitation commitments that can be applied toward the commitments of Annex I countries. It does not explicitly mention forest or land use but allows any project that has 'real, measurable, and long-term benefits related to the mitigation of climate change' and that is 'additional to any that would occur in the absence of the certified project activity.' The CDM allows the possibility of trading GHG mitigation from biomass energy promotion also, which may be fuel substitution or energy efficiency. Fossil fuel substitution by biomass fuel, i.e., electricity generation, biogas and biodiesel production from the biomass, may be important fuel substitution CDM projects in Bangladesh. Traditional biomass burning in Bangladesh is quite inefficient, so introduction of improved cooking stoves can be an important energy-efficient CDM project in the country. Technology upgrading for both the themes is critical to the CDM framework, which ultimately acquires the sustainable development of the country. Bangladesh, a Non-Annex I Party, ratified (accession) the Kyoto Protocol on October 22, 2001 (UNFCCC 2006c). So, Bangladesh is eligible to be a host country of CDM projects. Although forest management and conservation in Non-Annex I countries are not included in the first commitment period (2008–2012), they are likely to be included in the later commitment periods (Pearson et al. 2005).

## 3.3 General Overview of Bangladesh and Its Forestry Sector

Bangladesh is a South Asian least developed country located between 20°34′ to 26°38′ N latitude and 88°01′ to 92°42′ E longitude with a geographical coverage of 14.76 m ha with three broad categories of land hills, uplifted land blocks, and alluvial plains. The country is characterized by low per capita gross national product, low natural resource base, high population density, and high incidence of natural disasters. The climate is subtropical, characterized by high temperature, heavy rainfall, often excessive humidity, and fairly marked seasonal variations. Though more than half of the area is located in the north of the tropics, the effect of the Himalayan mountain chain makes the climate more or less tropical throughout the year (MoEF 2005). The country has an almost uniformly humid, warm, tropical climate. There are three main seasons: (1) a hot summer season, with high temperatures (5–10 days of more than 40°C maximum in the west), highest rate of evaporation, and erratic but heavy rainfall from March to June; (2) a hot and humid monsoon season (temperatures ranging from 20 to 36°C), with heavy rainfall from June to October (about two-thirds of the mean annual rainfall); and (3) a relatively cooler and drier winter from November to March (temperatures ranging from 8 to 15°C), when minimum temperature can fall below 5°C in the north, though frost is extremely rare (MoEF 2005). The mean annual rainfall varies widely within the country, according to geographical location, ranging from 1,200 mm in the extreme west to 5,800 mm in the east and northeast (MoEF 2005).

In 2004, the population of the country was about 141 m (USCB 2006) having 890 people per 100 ha in the country. Among the total population 79% lives in rural areas. The primarily agricultural economy of Bangladesh has recorded around 5% annual growth rate over the last few years (ADB 2001). The main crops grown in the country are rice and jute. Gross domestic product (GDP) was US $47,826 m in 2001 with the per capita GDP, US $364 (BB 2002).

Forestry is an important sector in Bangladesh's economy. It contributed about 1.84% of the country's GDP and 10.2% of the agriculture income in 2003/2004. The annual GDP of this sector in 2003/2004 was 4.48% (GOB 2004). Iftekhar (2006) reports that 'if environmental services and contribution in people's livelihood could have been properly accounted for, then the share of the forestry sector would have been much more.' Forestlands make up almost 18%, agricultural lands 64%, and urban areas 8% of the total lands in Bangladesh (FAO 1998). Other land uses account for the remainder. Total forestland area is 2.56 m ha, including officially classified and unclassified state lands, village forests, and tea/rubber gardens. Most of the state forestland is degraded. Classified and unclassified forestland signifies an administrative or legal category, not necessarily areas with forest cover. The natural forest accounts for about 31% and forest plantations 13% of total forest areas. Shifting cultivation, illegal occupation, and unproductive areas account for the remaining forestland (FAO 1998). Presently, protected areas represent just over 5% of forestland. The Bangladesh Forest Department is responsible for administering 65% of state forestland. The other government forestlands are administered by local District Commissioners (DC). The better quality natural forests and plantations in the government forestlands, excluding parks and sanctuaries (medium to good density), make up around 0.8 m ha, which is 5.8% of Bangladesh's total area. The area included in the present protected area network is 0.12 m ha, equal to 5.2% of state forestland or less than 1% of Bangladesh's total area (FAO 1998). On a per capita basis, there was 225 $m^2$ of forestland for every Bangladeshi citizen in 1993. This figure drops to 99 $m^2$ if only reasonably forested land, which excludes barren areas and low-density vegetation, is considered.

The hilly areas of Chittagong, the Chittagong Hill Tracts, Cox's Bazar, and the Sylhet Forest divisions, consist of hill forests, which are subject to severe degradation due to overpopulation, shifting cultivation, and extension of agriculture (Salam et al. 1999). In the hilly areas, two main types of forests are found, i.e., evergreen and deciduous. These forests may be subdivided into several subtypes based on altitude, soil, rainfall, and other factors. The evergreen forest is made up of tropical wet evergreen and tropical mixed evergreen. The deciduous forest consists of tropical moist deciduous and tropical open deciduous. Tropical mixed evergreen forest is the most important type, with the dominant tree species *Dipterocarpus* spp., which is highly valued due to its high-priced timber. In the forests of the hilly areas, more than 100 evergreen and deciduous tree species have been identified as growing naturally (Salam et al. 1999).

## 3.4 General Overview of the Republic of Korea and Its Forestry Sector

The Republic of Korea, a temperate country, is located at the heart of the Northwestern Pacific region. The Korean Peninsula encompasses 22.1 m ha, 45% of which (9.96 m ha) makes up the Republic of Korea ranging between 36°06′ to

38°27′ N latitude and 125°04′ to 131°52′ E longitude. The Republic of Korea has the world's 11th largest economy organized in the OECD (GoRok 2006). She took aggressive measures for economic development since the 1970s which prompted high year-on-year economic growth rate, attaining an average annual growth rate of 8.8% between 1986 and 1995. Successful international trading and industrialization expedited this development. The GDP rose 1.7-fold from US $252.5 billion in 1990 to US $427.3 billion in 2001 and per capita GDP rose 1.5-fold accordingly from US $5,890 in 1990 to US $9,025 in 2001 (GoRok 2006). Even though Korea stands as the world's third most densely populated country after Bangladesh and Taiwan, 474 per 100 ha in 2001, since the 1960s the population growth rate declined steadily from 3 to 0.71% in 2001, due to improved social and financial living standards, changed social perspective on population issues, and campaigns to control the growing population. The increase of GDP and steady decrease of population imply a fundamental economic change in the Republic of Korea. In the midst of 2006, the population of the country was 48.85 m (USCB 2006). Korea has a temperate climate with four distinct seasons. Winter is dry and cold due to the northwesterly wind sweeping down from Siberia. Korea has hotter summers and colder winters compared to the other countries located in the same latitude zone of the continent. Annual average temperature is 12–14°C in the central region and 3–10°C in the northern region. Annual mean rainfall ranges from 600 to 1,600 mm with uneven seasonal distribution (KFS 2006).

The Republic of Korea is a typical mountainous country where forest cover is about 64.3% (based on the 2004 forest statistics) to the total landmass covering about 6.4 m ha. The per capita forest is as low as 0.2 ha which is only one quarter of the world average. Total growing stock is 489 $mm^3$, with stock volume 76.4 $m^3$ $ha^{-1}$ (KFS 2006). Almost 59% of the forest stands are under the age of 30 years. Thus, Korea was largely dependent on imported timber, supplying about 94% of domestic timber consumption (KFS 2006). Forests in the Republic of Korea are classified into warm-temperate, cool-temperate, and boreal forests. Around 85% of them are identified as cool-temperate forests. Forest ecosystem is vulnerable to degradation due to soil conditions and heavy rains that occur during summer (Lee 2003; MoERoK 1997). During the Japanese Occupation (1910–1945) and the Korean War (1950–1953), illegal logging for forestry utilization resulted in a remarkable decline in the average stock volume from some 100 $m^3$ $ha^{-1}$ in early 1900 to 10 $m^3$ $ha^{-1}$ in 1960 (Lee 2003). To rehabilitate the degraded forestland, the Korean Government undertook a series of 10-year forest development plans including the massive reforestation programs. In 1999, a total of 97.4% of the degraded forestlands were rehabilitated very successfully by planting about 12 billion trees and raising growing stock to 70 $m^3$ $ha^{-1}$ in 2002 (Lee 2003). This is the largest reforestation success in comparison to the other countries in Asia. The increasing public demands for various forestry benefits as of watershed management and recreational sites have led the government to implement sustainable mountain development (MoERoK 1997). In the fourth Forest Development Plan undertaken in 1998, Korea laid the foundation for sustainable forest management by establishing more valuable forest resources, fostering competitive industries, and maintaining a healthy forest environment (Brown & Durst 2003). The main purpose of reforestation in Korea is to provide a range of environmental services along with timber products and diversification of farm income to the people (KFS 2006). Since 2001, sustainable forest management is being promoted through the basic forest legislation. Policies to increase sink in the forests are expected to be carried out through the forest tending projects, forest pest insect and disease management activities, forest fire management activities, mandatory replantation of harvested area, and promotion of urban greening (GoRoK 2006). Lee (2002) reported that the total accumulated carbon sequestration up to the year 2001 in the forests of the Republic of Korea was 14.4 m tC, with the annual carbon uptake rate of 1.83 tC $ha^{-1}$ $yr^{-1}$ in the degraded land and 2.14 tC $ha^{-1}$ $yr^{-1}$ in the shifting cultivation area reforestation, averaging 1.94 tC $ha^{-1}$ $yr^{-1}$. The carbon sequestration through the reforestation in the degraded land and shifting cultivation area shared 59 and 41%, respectively. Potential carbon sequestration up to 50 years from the establishment is expected to be 23.4 m tC in the country (Lee 2002).

The Republic of Korea is signatory to the UNFCCC and its subsequent extensions. As a Non-Annex I country, she ratified the Kyoto Protocol on November 8, 2002, thereby participating in the international GHG mitigation initiatives (UNFCCC 2006d).

### *3.4.1 What Bangladesh Can Learn from the Reforestation Success of the Republic of Korea*

During the Japanese colonization and the Korean War, the forest resources in Korea were diminished to a remnant from some 100 $m^3$ $ha^{-1}$ in early 1900 to 10 $m^3$ $ha^{-1}$ in 1960 (Lee 2003). Chun (2002) reports that Mongolian invasion in the 13th century and the subsequent colonization for 70 years destroyed the Korean forests severely. At that period 70% of the country was reported to be denuded. During the last 30 years starting from the 1970s, the Republic of Korea reforested almost all the degraded forestlands very effectively within a very shorter period in comparison to the other reforestation histories in the world (Chun 2002). Industrialization also advanced with the reforestation practice. Now the total growing stock is 489 $mm^3$, with a stock volume of 76.4 $m^3$ $ha^{-1}$ (KFS 2006). Chun (2002) reports that from the 1960s to the 1990s the growing stock was increased by 10 $m^3$ $ha^{-1}$ $decade^{-1}$. Nationwide reforestation, interests in creating forest resource base, forests as space for recreation and leisure, increased interests of citizen groups for forest conservation, increased concerns on land-use changes, and increased activism by citizen groups for social change made the effective and rapid reforestation success in the Republic of Korea (Chun 2002). People's participation and the positive attitudes toward reforestation acted as the main factor to make reforestation successful. Brown (2006) states that 'South Korea is in many ways a reforestation model for the rest of the world.'

The forests in Bangladesh faced degradation in different periods. During the British occupation (1757–1947), the forests were exploited to supply raw material for the ship and railway industries. The Pakistani colonial government also exploited the forest resources to earn revenue during 1947–1971 periods (Iftekhar 2006). Khan (2001) reports that the growing stock of wood was 21 $m^3$ $ha^{-1}$, with the productivity of 3 $m^3$ $ha^{-1}$ $yr^{-1}$ in 2001. Iftekhar (2006) estimates that if the forests continue to deplete in the current rate, the forest area will be less than 1 m ha and the growing stock will be only 9 $mm^3$, with 11 $m^3$ $ha^{-1}$ by 2050 if the other factors remain *ceteris paribus*. The government of Bangladesh took initiatives to rehabilitate the degraded forest areas through reforestation at different times. Now, reforestation through people's participation under the social forestry program has become a major activity of the Forest Department. Even though the reforestation projects were implemented with the grants from the funding organizations several times, the success of the plantation was very poor. Around 20–30% of all plantations established during the last 30 years have been destroyed and the growing stock of the remnants is less than satisfactory (Iftekhar 2006). Even community participation in the reforestation was not significantly successful (Ali 2002a). Considering the other developing countries' situation and the people's attitudes toward Bangladesh forest land use, Ali (2002a) stressed that the lack of success of conservation policies and a mismatch between forest policy and people's aspirations made the reforestation unsuccessful.

From the discussion of the reforestation trends and success in the Republic of Korea, it is clear that rapid economic growth and the people's understanding of the necessity of the forest resources/services expedited the success of the reforestation. Of course political commitment of the national leaders led the whole endeavor (Brown 2006). Figure 4.6 shows that Bangladesh has not achieved the fundamental economic development yet. In addition to this, the poverty rate is still about 50% in the country. Even though since the 1980s reforestation in Bangladesh was led by the people's participation, that participation was not spontaneous because of the lack of understanding of the people on the necessity of forest conservation and forest services. It can also be stated that the political commitment on the forest conservation and reforestation has not yet been created in Bangladesh.

Bangladesh can learn from the reforestation success of the Republic of Korea in different ways. Bangladesh can learn how the political leaders in the Republic of Korea created their commitment on the restoration of the degraded forestlands; how spontaneous mass participation was deployed for reforestation; how people's attitudes were altered from the mass collection of the forest products to the availing of the forest services; how the civil society were engaged to restore the degraded forestlands; and how the Republic of Korea made options of alternatives to the forest products. It is also important to figure out how the Republic of Korea matched reforestation with the advancement of industrialization. Finally, poverty reduction approaches and strategies during the last three to four decades of the Republic of Korea can be a great learning for Bangladesh. The learning from the

Republic of Korea can be of utmost importance for Bangladesh to be confident about the reforestation success which can move the country in track to have the CDM projects.

## 3.5 Potentials of the Bangladesh Forestry Sector to Mitigate Climate Change

The diversified ecosystems with the distinct type of plants act as an important carbon sink in Bangladesh. The total forestland is about 2.56 m ha including officially classified and unclassified state lands and forestlands in villages and tea/rubber gardens. More than half of the forestland is under different types of non-forest land use, e.g., shifting cultivation, illegal occupation, and unproductive areas.

In terms of per capita forestland, Bangladesh ranks among the lowest in the world, which is about 0.02 ha per person (ADB 1998). Throughout the country, most forestlands lack adequate natural cover, except negligible forest pockets. To conserve plants and other biodiversity, the Government of Bangladesh has declared a number of protected areas throughout the country. It covers only 5% of the existing forestlands. However, a vast majority of the land designated as forests is without tree cover. Most of the protected areas are not properly managed because of inadequate facilities and lack of proper implementation or enforcement of laws.

Most of the hill forests in Bangladesh are characterized as tropical evergreen or semi-evergreen forests. In such forests, the tropical evergreen plant communities are mixed with tropical deciduous trees, along with diverse herbs, shrubs, and bamboos. However, the natural looks and characters of the CHT forests have been changed due to severe human interventions.

The traditional *Sal* forest extends over Madhupur Tract, as well as the districts of Dhaka, Mymensingh, Rangpur, Dinajpur, and Rajshahi. In the *Sal* forest, 70–75% of the trees are *Sal* (*Shorea robusta*) (UNEP 2001). Most of the *Sal* forests have been denuded, degraded, and encroached upon by people or replaced by plantation of rubber monoculture and mostly exotic commercial fuelwood species.

Bangladesh has one of the most biologically resourceful and unique forests in the world known as the Sundarbans. It is the largest mangrove forest in the world (Iftekhar & Islam 2004). Out of 26 species of mangroves in the Sundarbans, the two dominant ones are the Sundri (*Heritiera fomes*) and Gewa (*Excoecaria agallocha*). At present, there is no commercial timber felling except Gewa and Goran (*Ceriops decandra*) due to a moratorium imposed by the Government of Bangladesh (pers. commun.).

The above scenarios reflect the fact that forestlands in Bangladesh can demand considerable payments for carbon credits, and this money may enhance developing the forestry sector in Bangladesh in many ways. Participation of Bangladesh in carbon trading will greatly augment the amelioration of the environment of Bangladesh as well as the world.

## 3.6 Implications of the Forestry Options for Different Land Uses

Different forestry options for different land categories should be targeted not only for sequestrating the carbon but also for meeting the biomass needs of local communities and industries; conserving soil, moisture and biodiversity; and generating employment for communities through the supply of non-timber forest products (NTFP). There can be four approaches increasing the carbon pool in the forests of Bangladesh: (a) conservation of forests and carbon sinks; (b) reforestation in previously forested barren lands and afforestation in newly accredited lands; (c) enrichment of the existing 'poor tree cover' forestlands with reforestation; and (d) enforcement of the forestry acts and regulations. All of these approaches are expected to achieve the objectives of forest resources development and abatement of GHG emissions. The following sections discuss the different forestry approaches and options to increase the carbon pool in the forests of Bangladesh.

### *3.6.1 Reducing Deforestation*

For forests with high standing biomass and low future growth rates, the best choice for the carbon emission reduction is simply to conserve the existing forest stand. Thus the 1.05 m ha of forestlands would fall in this category, simply requiring protection and enrichment (Table 3.1). These lands include medium to good

**Table 3.1** Forest land-use changes in legal public forestlands in Bangladesh

| | 1992 land use | | 1999 land use | |
|---|---|---|---|---|
| Type of land use | Area (m ha) | Percentage | Area (m ha) | Percentage |
| *Forestlands under forestry use* | | | | |
| Medium to good density natural forest areas | 0.46 | 20.85 | 0.20 | 9.15 |
| Protected areas other than in Sundarbans | 0.08 | 3.63 | 0.07 | 3.20 |
| Proposed protected areas: East Sundarbans | 0.03 | 1.36 | 0.02 | 0.91 |
| Proposed protected areas: West and South Sundarbans | | | 0.06 | 2.74 |
| Plantations in hills | 0.16 | 7.25 | 0.16 | 7.32 |
| Plantations in coastal areas | 0.11 | 4.99 | 0.08 | 3.66 |
| Plantations in protected and conservation areas | 0.006 | 0.27 | 0.006 | 0.27 |
| Other forest areas | 0.25 | 11.33 | 0.45 | 20.59 |
| Subtotal | 1.10 | | 1.05 | |
| *Forestlands not under forestry use* | | | | |
| Blank and encroachment | 1.11 | 50.32 | 1.14 | 52.15 |
| Total public forestlands | 2.21 | 100 | 2.19 | 100 |

density natural forest areas (0.20 m ha), established and proposed protected areas (0.15 m ha), plantations (0.25 m ha), and other forest areas (0.45 m ha). A total of 15 formally protected forest areas in Bangladesh occupy about 0.75% of the total land area of the country. The situation of these protected areas is quickly deteriorating as the pressure is being mounted from poaching, logging, and land conversion for shrimp farming. So, slowing down deforestation is the most important issue to preserve the forests and enhance the carbon sinks. The following sections discuss the causes of deforestation and how to slow down the deforestation rate.

#### 3.6.1.1 Causes of Deforestation and How to Slow Down This Process

Understanding the causes of deforestation is a prerequisite to any program to slow down the deforestation rate. Salam et al. (1999) indicate that deforestation and degradation in the forests in Bangladesh are influenced by infrastructural problems related to the country's underlying socio-economic features. Salam et al. (1999) divided the underlying factors into four sets of actors: (1) the indigenous forest dwellers, having their own problems (e.g., high population growth); (2) migrants, who move to the forests; (3) the timber industries cutting down too many trees; and (4) the government through its Forest Department which is not able or willing to implement suitable policies to regulate the cutting of trees and to prevent illegal cutting. Mitigating the first and second factors is a time-consuming task. The country is facing a severe problem of a high rate of population (Lush et al. 2000; Niroula & Thapa 2005). The constantly increasing population and its growing consumption expect further loss of forest cover due to these first and second factors. Thus, it is important to involve the forest dwellers in the forestry practices as much as possible. In contrast, the third and fourth actors can be seen as a relative indulgence. The nature of the causes of forest loss in Bangladesh is such that any attempt to revert these trends will be ineffective without changes in the attitudes and practices of the Forest Department officials and politicians with forest interests. The Forest Department has been losing its management capacity for many reasons, mostly related to the third and fourth actors.

The Natural Regeneration (NR) option is suggested for partially degraded forests; and Extended Natural Regeneration (ENR) is suggested for seriously degraded forestlands for the ecological role and to provide NTFP and small timber to the local communities. The role of NR and ENR is sketched in Fig. 3.1. It is suggested that tree felling for commercial purposes in forestlands should be totally banned. The NR and ENR will involve forest protection, soil and water conservation measures, and enhancement in planting of a few hundred stems of desired tree species per hectare. Ahmed et al. (1992) found the potential of natural regeneration of indigenous species in the degraded forestlands in Chittagong region of Bangladesh. Ahmed & Bhuiyan (1994) also found the natural regeneration capability in the Cox's Bazar forest areas of Bangladesh.

Conservation of existing forests would involve prevention of non-sustainable harvest of woody biomass meeting the timber needs from the private forestlands,

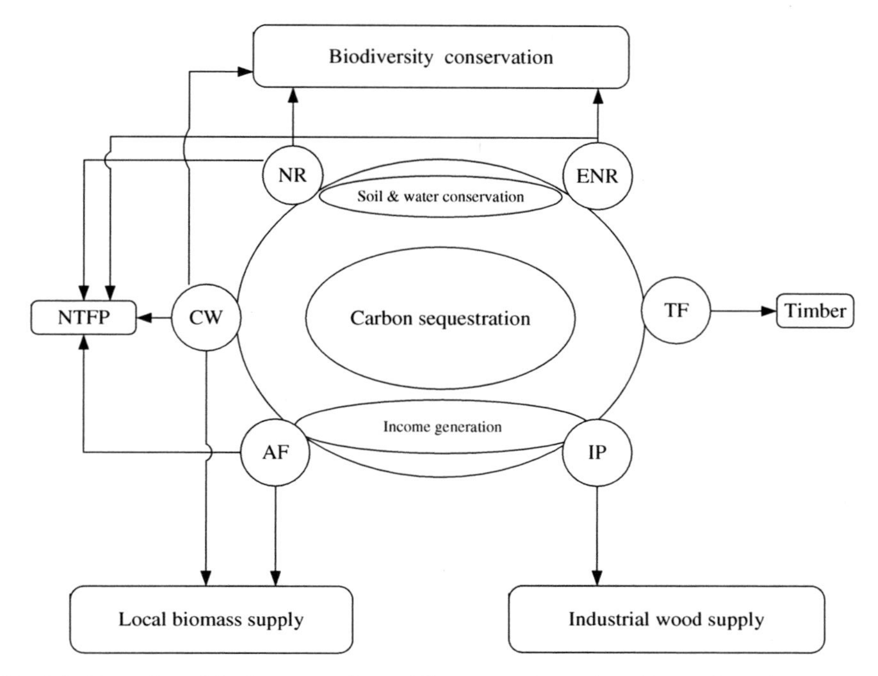


**Fig. 3.1** The utility of the forestry options while sequestering carbon. Notes: AF = Agroforestry; CW = Community Woodlot; ENR = Extended Natural Regeneration; IP = Industrial Plantation; NR = Natural Regeneration; NTFP = Non-Timber Forest Products; TF = Timber Forestry

banning access to natural forestlands by industries and commercial logging, encouraging private entrepreneurship in the forestry sector and biomass fuel conservation programs like biogas, and improved cooking stoves in rural and urban areas. The options considered in the next section aim to produce biomass to meet the needs of communities and industries, which would be fundamental to forest conservation.

### *3.6.2 Expansion of Carbon Sinks Through A/R*

Every year, the Forest Department takes up A/R programs for the whole country. Between 1974 and 1995, Bangladesh raised a total of 0.32 m ha of plantations of various types. Among these, hilly forests were 0.17 m ha, plain land *Sal* forests were 0.03 m ha, and coastal plantations were 0.13 m ha. By using a social forestry approach, the Forest Department is mobilizing a movement involving people from different communities to plant trees. A large number of trees have been planted in the community and private lands, e.g., on roadsides, embankments, school/institution premises, and canal banks. Presently, the afforestation rate per year is about 0.02 m ha (FAO 2000).

#### 3.6.2.1 Land Categories Available for A/R and Their Features

The potential for expansion of carbon sinks will be subject to the availability of land, land tenure and current vegetation status, and the cost of the land. The land categories available in Bangladesh and their features are discussed below. The identified categories represent the maximum potential land area available for increasing biomass cover.

Blank, Encroached, or Other Degraded National Forestlands

Table 3.1 shows that 1.14 m ha of forestlands under the control of the Government has been badly degraded or encroached upon by the local people (FAO 2000). Other forestlands are also being degraded. Khan (2001) reported that annual deforestation rate in Bangladesh is about 3.3%. The unclassed state forests are almost degraded to less than 10% crown coverage. These degraded lands need to be reforested with appropriate policies.

Homestead Lands

Homestead forests are spread over whole villages in the country with an efficient private ownership. These forests are usually called village forests not necessarily indicating the common pool resources. By providing a variety of goods and services, it acts as a buffer against the strong pressures on natural forests (Kumar & Nair 2004). It is reported that the homestead forests of Bangladesh provide about 70% of all wood consumed and 90% of all fuelwood and bamboo in the country (Alam et al. 1990). It was estimated that around 149 tree species exist in the village forests of Bangladesh with a total area of 0.27 m ha (FAO 2000; Khan & Alam 1996). Presently, around 70% of these forests have been degraded or declined due to the increasing population growth and fragmentation of the homestead lands. The spaces in the homestead lands may be planted with appropriate tree species, which can be managed by the carbon sequestration project.

Farmers with large holdings are more likely to keep their land fallow compared to the farmers with small or marginal holding (Alamgir et al. 2004; Miah & Hossain 2002; Shin et al. 2004). Thus by bringing the degraded and fallow land under tree plantations, employment opportunities will be generated for landless and marginal farmers in raising plantations and harvesting and processing of products. Large increases in biomass production in village ecosystems would improve the availability and access to biomass for the poor. Bangladesh has over 22.5 m bovine animals and 14.6 m goat/sheep only for 14.12 m ha of the total land mass (BBS 1996). Thus for any forestry option in community and forestlands or even the degraded private lands, the grass/fodder or grazing requirements require suitable silvicultural practices to ensure fodder growth. Given the current low grass productivity of degraded lands it will be possible to ensure higher fodder yields through A/R.

### *3.6.3 Different Forestry Options for A/R*

Different forestry options are considered for different categories of lands based on current ownership and usage and the potential to deliver benefits to communities and environment. The features and potential benefits of different options are summarized in Fig. 3.1.

#### 3.6.3.1 Community Woodlot (CW)

Khas lands,[1] where available, should be dedicated to Community Woodlot (CW) to primarily meet the diverse biomass needs of local communities in a sustainable way. Thus trees meant for timber and NTFP would lead to sequester carbon. Only the sustainable harvest of fuelwood, NTFP, poles, and timber would be permitted.

#### 3.6.3.2 Timber Forestry (TF)

All the area under miscellaneous tree crops and a part of cultivatable wasteland can be considered for TF, which would be adequate to meet the projected demand. TF aims to meet structural timber needs, where the wood has a long life and stores carbon for long periods.

#### 3.6.3.3 Industrial Plantation (IP)

Industrial plantation aims to meet the requirement of the large-scale forest-based industries in the country, i.e., pulp and paper, hardboard, and match industries. Only the fully degraded national lands can be considered for this type of plantation. Carbon sequestration takes place in three ways: first, the soil carbon storage will increase; second, at any time, some fixed quantities for biomass will be present in the vegetation; and third, and more importantly, growing biomass for industrial uses in degraded lands will lead to conservation of existing forest carbon sinks, which would have been depleted to meet the softwood needs.

#### 3.6.3.4 Agroforestry (AF)

All the rainfed or dry farming croplands can be considered for AF. These croplands are usually private lands in Bangladesh owned by the rural farmers. Carbon sequestration would take place in two ways: first, trees grown for timber and NTFP would sequester carbon and second, growing biomass on farm lands will lead to reduction in pressure on forest biomass.

For the CW and AF, silvicultural practices should be based on local indigenous knowledge. Species selection also should be left to the farmer's choice.

---

[1] Khas lands in Bangladesh are defined as the degraded fallow lands owned by the Government.

### *3.6.4 Enforcement of the Forestry Acts and Regulations and Boosting Up the Responsibility as a Member of the International Conventions*

Bangladesh has signed the Convention on International Trade in Endangered Species of Wild Fauna and Flora (CITES) in 1973, the UNFCCC in 1992, the Convention on Biological Diversity (CBD) in 1992, and the Kyoto Protocol in 2001. She is a signatory to the Ramsar Convention and the World Heritage Convention. The Bangladesh Wildlife (preservation) Act, 1974, the Forest Act, 1927 (amended in 1989), the Fish Act, 1950, and the Environment Protection Act, 1995, provide legal support for forest and biodiversity conservation in Bangladesh. The national forest policy, 1994, also supports the mass reforestation activities throughout the country. Bangladesh already has adequate legislation and administrative machineries to supervise the forest areas. However, what is crucial is the need to recognize that the biomass needs of the local communities can be and should be met from the forests in a sustainable way through community participation and sustainable management. Conservation of existing forests would involve prevention of non-sustainable harvest of woody biomass, meeting the timber needs from the private forestlands, banning access to natural forestlands for industries and commercial logging, encouraging private entrepreneurship in the forest sector, and biomass fuel conservation programs like biogas and improved cooking stoves in rural and urban areas. To mitigate global warming through the forestry sector in Bangladesh, it is necessary to adjust or pinpoint the objectives of the forest policy which should be compliant with the biodiversity conservation in the forests.

## 3.7 Issues to Be Settled for Carbon Credits

### *3.7.1 Crediting*

Due to the leakage and non-permanence risks in the 'LULUCF' projects, parties agreed that credits arising from CDM-A/R projects should be temporary, but could be reissued or renewed every 5 years after an independent verification to confirm that sufficient carbon was still sequestered within the project to account for all credits issued (Pearson et al. 2005). This is also assumed to be more flexible to the Non-Annex I countries in comparison to long-term crediting system (Võhringer 2004). In the temporary crediting, carbon accounting should be implemented on a carbon tonne-year basis. Under a tonne-year system, credit would be given for the number of tonnes of carbon held out of the atmosphere each year. Under the tonne-year accounting system, delaying deforestation merits credit irrespective of the long-term fate of the forest, although the cumulative credit that can be earned from a given patch of forest is obviously greater as the forest remains standing longer.

### *3.7.2 Protected Area Establishment and Crediting*

Based on the current criteria of CDM, establishing a park in an area of forest that would not be cleared receives no credit, whereas one in an area experiencing rapid clearing is heavily rewarded (Fearnside 1999). The park in the area with little clearing is likely to be cheaper to establish. How to allot carbon credits can, therefore, influence where to create parks.

In Bangladesh, the least protected and most threatened types of forest are in the Central and Southeastern part of Bangladesh where reserves may be established with little clearing (Salam et al. 1999, 2005).

### *3.7.3 Reduction of Fossil Fuel Use or Maintenance of the Stock of Forests*

Creating and maintaining carbon stocks in the forests is an important climate change response option for Bangladesh. Although carbon stock maintenance through forest management and conservation in Non-Annex I countries is not included in the first commitment period, it is likely to be included in later commitment periods, but strong arguments are made regarding incorporating this form of environmental service into global warming mitigation policies.

If carbon stock maintenance is recognized as a form of mitigation measure, as distinguished from avoided deforestation, then monitoring needs would be much simpler from the point of view of countries contributing funds as carbon credits; only accompaniment of forest stock remaining each year would be necessary (Fearnside 1999). Even though fossil fuel reduction is generally assumed to be much easier than maintaining the carbon stock in the forests, the value of forests for climatic functions other than carbon stocks and for maintaining biodiversity and indigenous cultures provides additional reasons to treat them differently from fossil fuel reserves (Meng et al. 2003). Bangladesh as one of the least developed countries has not yet developed the appropriate biomass-based technology to substitute the fossil fuel and it is not likely to be achieved in the near future, so carbon stock creation and maintenance in the forests may be the relatively beneficial option (Baral & Guha 2004; Huq 2001).

### *3.7.4 Internal Carbon Credit Allocation Regimes*

Under the Kyoto Protocol, credits earned for sequestering $CO_2$ would be allocated to the governments that have ratified the international treaty. The governments may allocate these credits and Kyoto obligations to domestic forest owners or retain them as they see fit. Since there is still much uncertainty over the mechanics of how credits will be allocated to growers, Guthrie & Kumareswaran (2003) introduced three alternative schemes, i.e., lump sum regime, the flows regime, and the stocks regime. Among them the lump sum regime and the stocks regime seem to be appropriate for Bangladesh considering the existing socio-economic conditions. The lump sum regime entitles forest owners to a one-off lump sum payment for engaging in A/R, paid at the commencement of the first rotation. Growers retain these credits as long as replanting occurs immediately after harvest; if the forest is not replanted, deforestation has occurred, and the owner incurs an obligation, the lump sum allocation must be repaid in full. The stocks scheme, which is the one modeled by Sohngen & Mendelsohn (2003), rewards this temporary storage by giving forest owners credit not only for the quantity of carbon captured but also for the duration of capture, effectively treating carbon credits as rent payments for the storage of carbon. Under this regime, credits are allocated to the owner annually in proportion to the size of the total carbon stock, which grows cumulatively through time. In order to increase carbon sequestration, forest owners must be encouraged to lengthen rotations and be discouraged to convert forestland to alternative uses.

## 3.8 Conclusions

The discussion on the forest resources in Bangladesh indicates that forestlands are potential to sequester carbon to reduce the global warming. The chapter emphasizes both conservation of forests and carbon sinks and expansion of carbon sinks through A/R. Meeting the biomass needs of the people has been discussed, which can be regulated by sustainable harvesting except in the protected areas. This study has figured out different forestry options aimed at carbon sequestration with other necessary utilities. But the unsettled issues discussed should be settled along with appropriate policies indicated in this study. Policy changes are expected to have the greatest potential effect in this arena. As slowing down the deforestation is an important global warming response option that can gain potential carbon benefits as expected by the CDM in the second commitment period of the Kyoto Protocol, it is imperative that needed efforts should be made to develop this option by the Parties. A/R, which is much closer to offering eligible projects for investment, has the principal barriers, which are social in most cases in Bangladesh. Steps should be undertaken to ensure that unacceptable social impacts do not derive from the plantation expansion programs. The discussion on the reforestation success in the Republic of Korea made clear that rapid poverty alleviation, spontaneous mass participation, and political commitment acted as a mainstream to reforest the degraded forestlands effectively. Bangladesh can make an intergovernmental collaboration with the government of the Republic of Korea to learn how a reforestation scheme can be successful. This study is expected to play a great role in mitigating global climate change through the different forestry options in Bangladesh.

# Chapter 4
# Implications of Biomass Energy and Traditional Burning Technology in Bangladesh

**Abstract** Globally, biomass contributes to the largest share to primary energy supply in developing countries, Bangladesh in particular. The major use of this biomass energy is for domestic cooking in the rural areas. As the climate change mitigation response option, the sustainable production of biomass and its supply for primary and secondary energy along with the improvement in the traditional burning device, can be important CDM activities in Bangladesh. This book reckons the energy resources and its consumption in Bangladesh. It attempts to discuss on the potentials and problems of forest biomass productions and how it can be incorporated into the CDM projects in Bangladesh. It finds biomass energy based CDM projects potential for accruing sustainable development in Bangladesh. This findings and discussion in this chapter can be important input tor the renewable energy policy in Bangladesh.

## 4.1 Introduction

Due to the non-renewability of the fossil fuels and its foreseeable depletion, a close attention is being given to biomass energy for its renewability and potentials for climate change mitigation (Bhattacharya et al., 2003). Biomass energy stands as an attractive option, able to both satisfy the socio-economic requirements imposed on CDM projects and contribute to climate change mitigation objectives (Silveira 2005). In particular, wood fuel holds great promise as a component of CDM strategies that are meant to operationalize international agreement to reduce $CO_2$ emissions to acceptable levels. Furthermore, extensive establishment of forest energy plantation could provide livelihood opportunities to impoverished rural communities in developing countries. Biomass grown in deforested land and degraded areas helps to restore the biodiversity of the system and also to prevent soil erosion. Biomass energy systems are not site specific, these can be established in any place where plants are grown and animal wastes are available. Of these renewables, biomass has an added advantage of being able to be set up on a small scale to provide power and electricity to villages and small cluster or on a large scale for electrical power generation to be fed to the national grid. Thus, there is a need to produce woody biomass not only as fuels but also as a means to address climate

Md.D. Miah et al., *Forests to Climate Change Mitigation*, Environmental Science and Engineering, DOI 10.1007/978-3-642-13253-7_4, © Springer-Verlag Berlin Heidelberg 2011

change-related issues and socio-economic problems. If adequate policy initiatives are provided, by 2025, 30% of the direct fuel use and 60% of global electricity supplies are expected to be met by renewable energy sources (Koh & Hoi 2003).

Developing countries meet the energy needs mainly by traditional biomass fuels (38%) (Ravindranath et al. 2006; Sims 2003). In some countries such as Bangladesh, Kenya, and Paraguay, its use is as high as 75–90% (UNEP 1994). Traditional biomass use, mainly for cooking and heating, is characterized by low efficiency of use. The unsustainable extraction and use of traditional biomass energy leads to degradation of the local environment and forests, deforestation and the consequent loss of forest products, soil erosion and loss of biodiversity, and domestic air pollution affecting human health. But the modern forms of biomass energy provide numerous environmental benefits.

Bhattacharya et al. (2003) for some selected Asian countries; Elauria et al. (2003) for the Philippines; Koh & Hoi (2003) for Malaysia; Perera et al. (2003) for Sri Lanka; Sajjakulnukit & Verapong (2003) for Thailand; Sudha et al. (2003), Ramachandra et al. (2000), and Sudha & Ravindranath (1999) for India emphasized the importance of sustainable biomass promotion in the perspective of global climate change. Berndes et al. (2003) focused on the contribution of biomass in the future global energy supply. Baral & Guha (2004) and Marland & Schlamadinger (1997) showed the logic to replace the fossil energy by biomass energy.

The energy requirement of Bangladesh is met from different sources, such as biomass, electricity, natural gas, kerosene, diesel/gas oil, and coal. The people of rural areas depend mainly on traditional fuels, namely wood, and agricultural residues, such as paddy husk, bran, straw, bagasse, jute stick, twigs, leaves, fuelwood, charcoal, and cowdung for their domestic consumption. Among the biomass, tree and bamboo provide 48%, agricultural residues 36%, and dung 13% of the current domestic energy requirement. The remaining 3% is supplied from other sources (GOB 1993).

In Bangladesh, about 85% of the population lives in the rural areas in 15.61 m households spread over 68,000 villages, with an average family size of 5.6 per household (BBS 2000). The chief sources of energy used in Bangladesh are biomass fuel, natural gas, oil, coal, and electricity. In the rural areas in Bangladesh, foods are cooked in the traditional cooking stove where biomass fuels are mainly used as energy because of the unavailability of natural gas. It is assumed that large quantities of GHGs are emitted from these stoves because of the inefficiency of biomass burning leading to the wastage of the burning capacity of the wood fuel. This also leads to increased collection of wood fuel from village forests and consequent deforestation as the production of fuelwood in the forests is not sustainable. The design of the traditional cooking stove and the type of fuel used are important determinants of fuel use efficiency, which is related to the emission of GHGs (Bhattacharya et al. 2002). Greenhouse gas emission reduction requires efficient use of wood fuel. Optimum efficiency of wood fuel use can reduce the emission of GHGs to the atmosphere as well as the deforestation of village forests.

In Bangladesh, fuelwood was considered a free product in the rural areas in the 1950s. Now, the shortage of fuelwood is severe throughout the whole country due to the overexploitation of natural forests and village groves, which has made the fuelwood as market commodity at present time. The Forestry Master Plan (FMP) of Bangladesh (GOB 1993) states that per capita fuelwood consumption in Bangladesh is the lowest in the world. The projection of fuelwood demand and supply for the years 1993–2013 by the FMP shows a disappointing figure of the fuelwood deficit. This shortage will also be severe if the country fails to protect the forest cover and develop the fuelwood plantation throughout the country. So, forest biomass production options should be searched out in both public and private forestlands in the country. These production and conversion to energy under the framework of CDM are important. Biswas et al. (2001) urged that if sustainable development is to occur in Bangladesh and similar countries, improvements are required in the energy sector. Studies on biomass fuel use were carried out in different regions of Bangladesh. Koopmans (2005) described the overall situation of wood fuel production and supply in Bangladesh including the other South and Southeast Asian countries. Abedin & Quddus (1990), Alam et al. (1990), Dasgupta et al. (1990), Kar et al. (1990), and Miah et al. (1990) studied the household biomass energy situation in the forest-poor regions of Bangladesh. Akhter et al. (1999) studied the homestead biomass fuel energy situation of a forest-rich district, Cox's Bazar. No study was so far carried out to show the prospects and to find out the options of biomass energy promotion under the framework of CDM. This chapter deals with the forest biomass production options and figures out the traditional cooking stove issues in relation to GHG emission and deforestation. It also discusses on the prospective roles of CDM to promote biomass energy in Bangladesh.

## 4.2 Energy Resources and Consumption

The main energy sources of Bangladesh are biomass, oil and coal, and natural gas. Biomass energy sources are traditionally used for domestic cooking and in small rural industries. Biomass fuels are estimated to account for about 73% of the country's primary energy supply (WB 1998) as shown in Fig. 4.1. Bangladesh has proven natural gas reserves of 301 b $m^3$ that would last for 32 years at the current rate of production. The remainder of energy supply is from oil, mostly imported, and limited amount of hydropower. Bangladesh has an installed electric generating capacity of 4,005 MW, of which 94% is thermal, mainly natural gas fired (BPDB 2002).

The country's per capita annual energy consumption in 1997 was about 77 kgoe, and it was much below the world average of 1,474 kgoe (ADB 2001). Only around

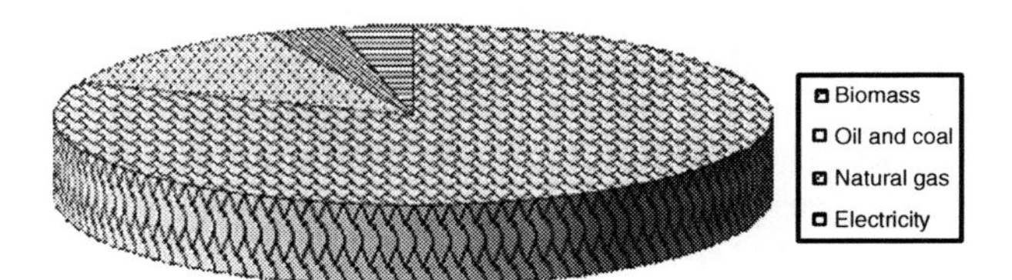


**Fig. 4.1** National energy consumption in Bangladesh

30% of the population has access to electricity (BPDB 2002). Over 80% of the people depend on traditional energy sources such as firewood, cow dung, and agricultural residues for their energy needs. Excessive usage of firewood threatens the remaining forest cover, which is only 10% of the total land area (WEC 2000).

## 4.3 Present Forest Land Use and Its Problems

The total forestland area equals 2.56 m ha, including state forest, village forest, and tea/rubber gardens. More than 90% of the state-owned forestland is concentrated in 12 districts in the eastern regions. Out of 64 districts in the country, there is no national forestland in 28 districts (FAO 1998). So, the national forestland in Bangladesh is eccentric. Figure 4.2 shows the three main types of forest cover in Bangladesh.

About 6,000 ha of forestland was believed to be lost each year due to various reasons. But recent observations by the Forest Resource Assessment Project suggest that the annual rate of forest destruction has exceeded 37,600 ha (FAO 2000).

Ali (2002a) describes that illegal harvesting, encroachment, and shifting cultivation are the important problems in forest land use in Bangladesh. In Bangladesh, individual offences related to illegal harvesting are mostly for fuelwood, home, and farm implements or for the sale of goods in the market for personal daily livelihood requirements. Although per capita fuelwood consumption of Bangladesh is one of the lowest in the subcontinent, fuelwood supply from the state forests of Bangladesh has not been enough to meet the demand of the whole population (BBS 1997). Official fuelwood supply from the forest (1985–1990) was only 0.7 $mm^3$ $yr^{-1}$, whereas the demand was about 7.0 $mm^3$. The case for timber was similar. The average supply from the forest was 1.09 $mm^3$, whereas the demand was 2.42 $mm^3$. As the gap between supply and demand continued to grow, illegal harvesting was reported to be increased. The comparative per capita consumption of fuelwood and timber in some selected Asian countries has been given in Fig. 4.3 adopted from Ali (2002a). As a result of forest clearance, encroachment is one of the worst problems of Bangladesh forest land use. Once the forest is cleared and left unforested, people start to invade and claim the land as their own. Ali (2002a) describes that about 62,000 ha of national forestland has been encroached up to December 1980, and more than 88,000 people were living inside forests. Moreover, there were about 5,000 forest villagers living inside forests legally, whose number was also increasing.

Shifting cultivation is an age-old practice in the hilly areas of Bangladesh. Although it is not the common practice of plain-land people, the way tribal people conduct shifting cultivation, particularly through the use of fire and terracing of hilly land, was considered highly detrimental to the forest environment (Ali 2002a). According to GOB (1993) about 60,000 households were practicing shifting cultivation at that time on 85,000 ha of land with an average of only 1.3 ha land per household.

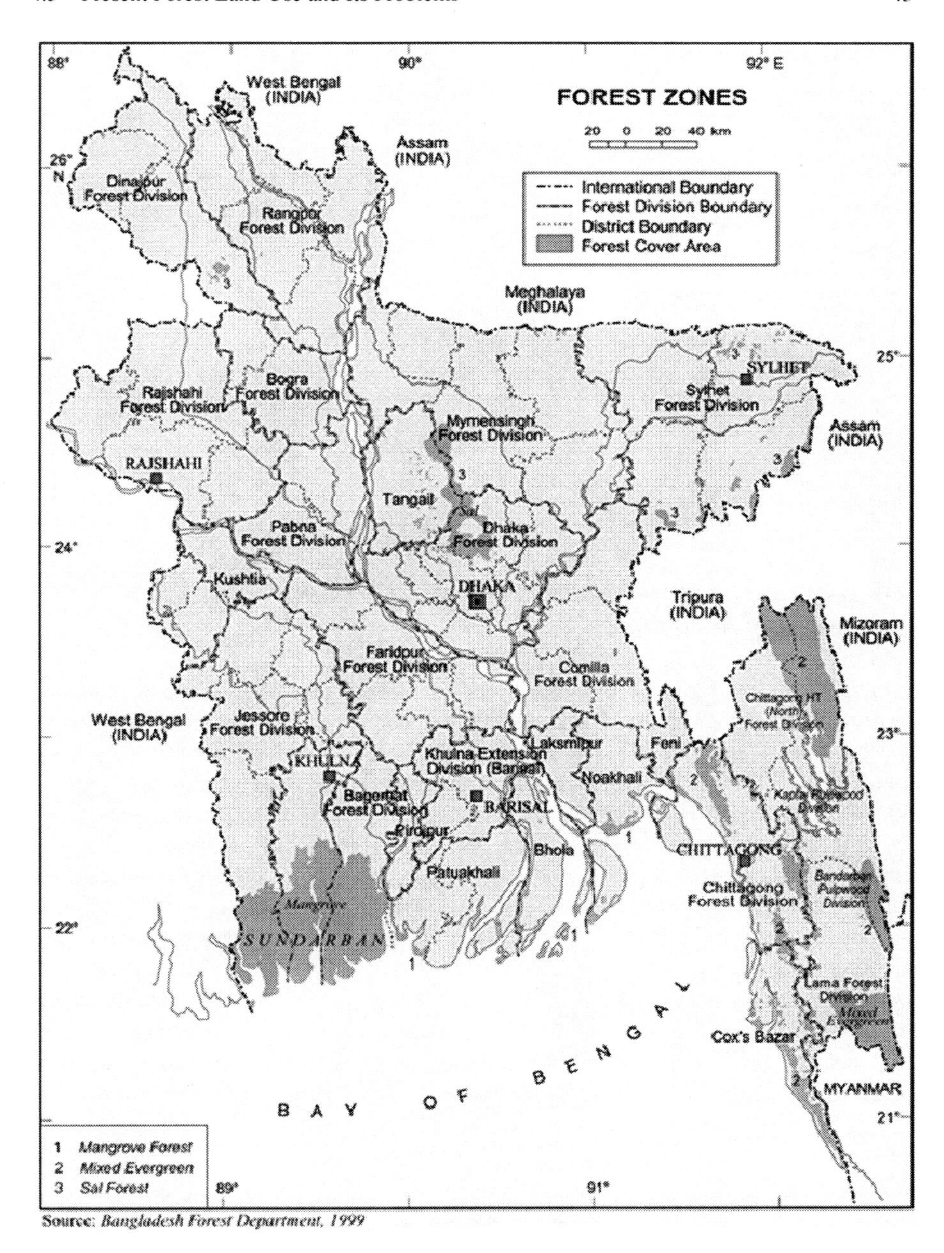


**Fig. 4.2** Major three forest types in Bangladesh

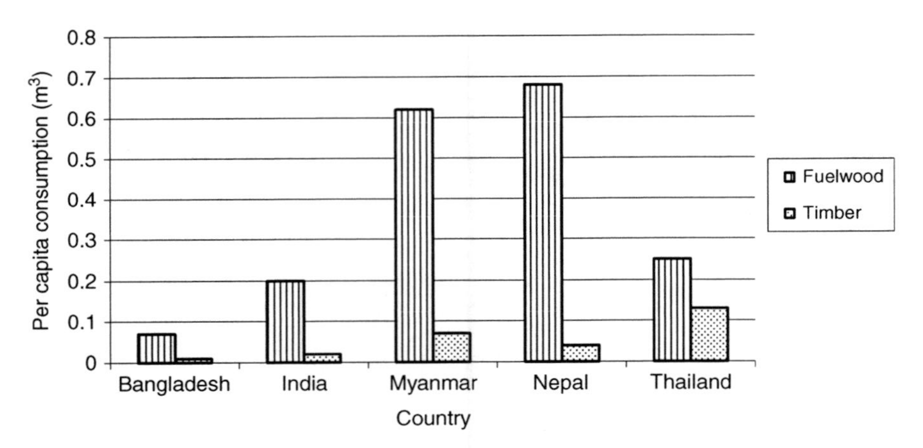


**Fig. 4.3** Per capita consumption of fuelwood and timber in some selected Asian countries

## 4.4 Biomass Demands on Forests and Different Forestry Scenarios

Forests provide a range of biomass products to rural and urban communities and industries. The population of the country has leapt from 45.646 m in 1950 to its present figure of about 141.340 m and is predicted to reach 219.636 m by 2030 and 279.955 m by 2050 (USCB 2006). Thus, it is imperative to follow the strategy adopted by energy analysis, which is to aim at reducing carbon emissions while ensuring an adequate supply of energy services to meet the development needs of a country. FAO (2000) has projected biomass demands for wood-based raw materials for 55 years starting from 1995 (Table 4.1). The first scenario can be considered as a baseline scenario, under which there is no further new forest plantation establishment in the future. This scenario does, however, assume that all harvested areas will be replanted with the same species and for the same purposes (i.e., industrial roundwood or wood fuel production) in the year that they are harvested.

Thus, the total area of forest plantations remains constant throughout the projection period (to 2050). Variations in projected potential roundwood production

**Table 4.1** Projected potential roundwood production from plantation (1995–2050) in Bangladesh

| | Industrial roundwood (thousand $m^3$) | | | | | Wood fuel (thousand $m^3$) | | | | |
|---|---|---|---|---|---|---|---|---|---|---|
| Scenarios | 1995 | 2000 | 2010 | 2020 | 2050 | 1995 | 2000 | 2010 | 2020 | 2050 |
| Scenario I | 386 | 477 | 512 | 540 | 492 | 35 | 151 | 669 | 417 | 590 |
| Scenario II | 386 | 477 | 512 | 611 | 703 | 35 | 151 | 678 | 495 | 823 |
| Scenario III | 386 | 477 | 512 | 762 | 930 | 35 | 151 | 698 | 657 | 1,082 |

over time are, consequently, the result of changes in the forest plantation age-class distributions and for different species.

A second 'medium growth' scenario was modeled under the assumption that forest plantation areas would increase each year by an amount equal to 1% of the forest plantation area in 1995. Again, this scenario also assumes that all harvested areas will be replanted with the same species and for the same purposes in the year that they are harvested. It is also assumed that the species used and the purpose of newly planted areas will be in proportion to the current species mix and use, such that these variables do not change relative to each other in the future. Projected potential production under this scenario starts at the same level as in Scenario I, but increases to a higher level due to the expansion of forest plantation areas managed on short rotations and, later on, as other newly planted areas reach maturity.

The third scenario in this analysis contained the highest assumption about future rates of new planting. Again, this scenario also assumes that all harvested areas will be replanted with the same species and for the same purposes (i.e., industrial roundwood or wood fuel production) in the year that they are harvested. It is also assumed that the species used and the purpose of newly planted areas will be in proportion to the current species mix and use, such that these variables do not change relative to each other in the future.

The doubling or trebling of demand for biomass-based products is likely to lead to increased pressure on forests and large-scale imports.

## 4.5 Land Categories Available for Biomass Production and Their Features

The potential for expansion of biomass plantation will be subject to the availability of land, land tenure and current vegetation status, and the opportunity cost of land. The land categories available in Bangladesh for this purpose and their features have been discussed in Sect. 3.6.2.1. The identified categories represent the maximum potential land area available for increasing biomass cover.

### *4.5.1 Different Forestry Options*

Bangladesh has been implementing a large plantation program since 1974. Up to 1995, she had developed 0.32 m ha of plantation through A/R activities. At an afforestation rate of 18,000 $yr^{-1}$, over 60 years are required to afforest the 1.14 m ha of the degraded area.

Different forestry options are considered for different categories of lands based on current ownership and usage and potential to deliver benefits to community and environment. The features and potential benefits of different options for biomass production through A/R are summarized in Sect. 3.6.3. It also has been visualized in Fig. 4.4.

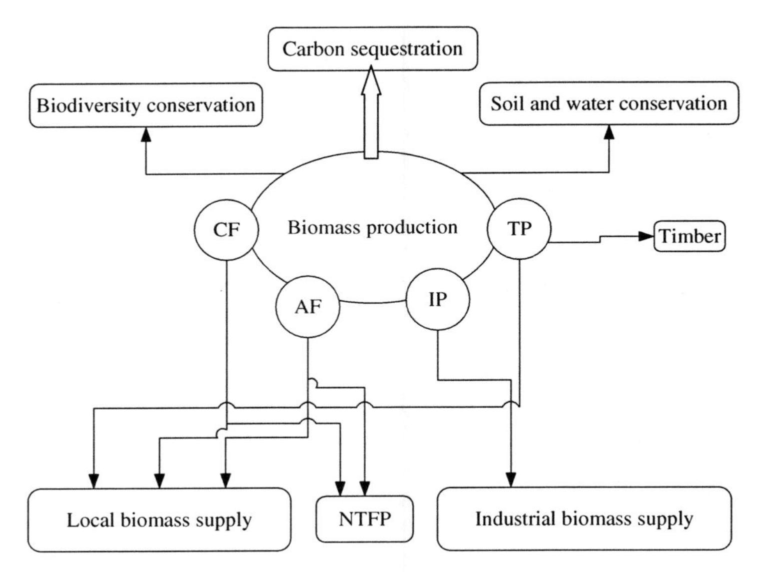


**Fig. 4.4** The utility of the forestry options while producing biomass in the forestlands
Notes: AF = Agroforestry; CF = Community Forestry; IP = Industrial Plantation; NTFP = Non-Timber Forest Products; TP = Timber Plantation

## 4.6 CDM Additionality for Biomass Energy Promotion in Bangladesh

Bangladesh is a poor developing country in South Asia. Major national problems in Bangladesh are increasing population and poverty (FAO 1998). Figure 4.5 shows the annual growth trend of population and GDP from 1985 to 2003 (SESRTCIC 2006). In 2003, the annual growth of population had dwindled to 1.37% from 2.22% in 1985. The GDP growth had reached 5.3% in 2003 from 3.2% in 1985 (ADB 2004). The decrease of growth of population and the increase of GDP seem to be so slow, which implies that fundamental economic change has not occurred considering the population change. In addition to this, still Bangladesh has poverty (under the National poverty line) of 49.8% (ADB 2004). In 1984, 1989, and 1992, the poverty rates were 52.3, 47.8 and 49.7%, respectively (Sen 2003). The expected biomass energy promotion in Bangladesh includes briquette, charcoal, biogas, and biodiesel production; electricity generation by biomass; and introduction of improved cooking stove to the household level. Among them, charcoal from wood and briquette from agricultural and wood residues are being produced with very primitive technology in the country. With some funded program, the government and NGOs are promoting biogas in a very limited scale. The modern technology transfer for the ongoing biomass energy production is simply impossible for Bangladesh government due to severe poverty. Likewise the rest of the biomass energy promotion is not possible in Bangladesh with her own fund and technology. The important

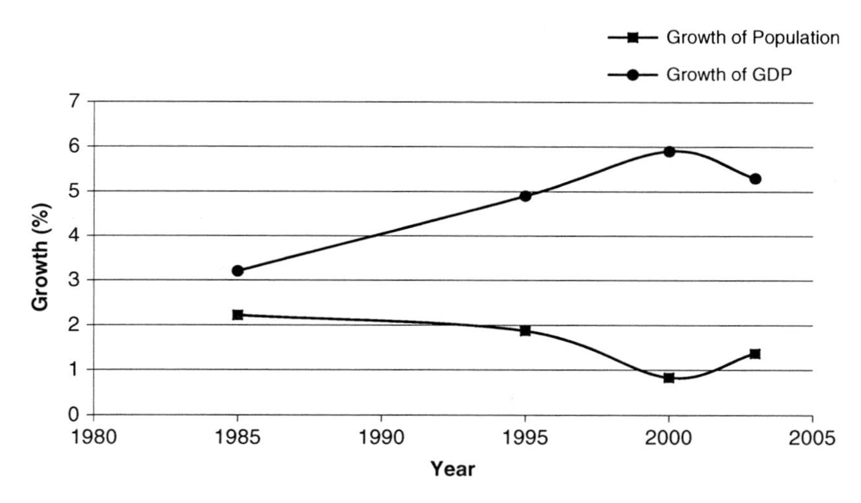


**Fig. 4.5** Annual growth of population and GDP in Bangladesh

barriers to biomass energy promotion in Bangladesh may be categorized into technical, financial, institutional, and policy. The technical barrier to biomass production for energy includes low biomass productivity and lack of experience in emerging biomass conversion technologies. The main financial barrier to biomass production for energy is the high-energy production cost, difficulty in accessing finance, and unavailability of incentives. It also includes shortage of investment in the forestry sector. The institutional barrier includes lack of coordination among different government agencies and lack of mechanism for their interaction on a regular basis with private sector, lack of a designated agency for promoting biomass energy/plantation, and lack of access to expertise on plantation in degraded land as well as extension service to potential planters. The policy barrier includes unclear, unsupportive, and biased (in favor of conventional energy sources) government policy and absence of national strategy or priority given to biomass for energy use. This situation shows the clear scenarios of 'additionality' of biomass energy promotion in Bangladesh under the framework of CDM. So, only CDM projects can promote the biomass energy in Bangladesh and achieve the 'certified emission reduction' and contribute to the sustainable development of the country.

## 4.7 Traditional Cooking Stove Issues in Bangladesh

Cooking energy constitutes a major fraction of energy consumption in rural areas of developing countries which is largely met with biofuels, such as fuelwood, charcoal, agri-residue, and dung cake (Rubab & Kandpal 1996). Majority of the households (90%) in Bangladesh cook foods and produce heat for other purposes by the traditional stoves (Hossain 2003). 'A traditional stove is usually a mud-built cylinder with three raised points on which the cooking utensil rests' (Hossain 2003). Among these raised points, one opening is used for the fuel feeding and the other two openings for exiting flue gases. In some cases, single cooking stoves (only one flue gas exit and another fuel feeder) are also made. The traditional cooking stoves either can be made underground or aboveground. Hossain (2003) comments that the traditional cooking stoves used in Bangladesh allows so much loss of heat due to the large distance between the pot and the fuel bed (depth ranging from 30 to 60 cm); because of the large size of flue gas exits between the cooking pot and stove, air cannot reach the bottom of the stove. Traditional biomass-fired stoves have been identified as inefficient due to its incomplete combustion system (Qiu et al. 1996; Rubab & Kandpal 1996). The thermal efficiencies of traditional cooking stoves vary from 5 to 15%, depending on the depth of the stove and the size of flue gas exits (Khan et al. 1995). Bhattacharya & Salam (2002) and Streets & Waldhoff (1999) report that it causes significant GHG emissions due to formation of products of incomplete combustion and poses health hazards. Qiu et al. (1996) also report that due to the unsustainable biomass production, it causes ecological and environmental problems, such as deforestation and land degradation.

Due to the inefficient use of wood fuel throughout the country, widespread destruction of forests including homestead forests has reduced the forest cover to about 8% from the minimum requirement of 25% of the land area of the country (ADB 1998). Collection of fuelwood for use in the traditional cooking stove is not the sole cause of deforestation in the homestead forests of Bangladesh. As the national forests in Bangladesh have been significantly degraded, they are no longer able to fully supply the needs of the people of the country for forest products. So, for timber, bamboo, and other non-timber forest products, the people are mostly dependent on homestead forests, which have their implications with the recent deforestation in the rural areas. But in terms of GHG emissions, using traditional cooking stoves has some serious implications. It is an already proven fact that traditional cooking stoves in rural Bangladesh are inefficient due to the incomplete combustion of the fuelwood. This low efficiency results in high consumption of fuelwood which leads to more collection of fuelwood from the forests. Furthermore, plantations in rural areas are not sustainable and so are not able to contribute to net carbon sequestration. The major GHGs emitted from traditional cooking stoves, $CO_2$, and $CH_4$ (Ali 2002b; Bhattacharya & Salam 2002; Smith et al. 2000), have already been proved as the primary GHGs associated with the global warming potential (IPCC 2001b).

The tree species used for fuelwood may vary in different agro-ecological zones in Bangladesh, but the tendency has been shown as the same for the sources of collection. Miah et al. (2003) found that among the various sources used by households for collecting fuelwood, homestead forests were the most important (33%), followed by markets (28%), secondary forests/plantations (25%), and agricultural residues (15%). This shows the same trend for sources of fuelwood in all ecological zones in Bangladesh differing with secondary forests/plantations.

Because of incomplete combustion of biomass fuels in traditional cooking stoves, some irritants, toxins, and carcinogens are released in the kitchen environment and these pose a major threat to the respiratory system of the users (Ezzati et al. 2004; Mishra et al. 2005a, b; Smith 2002a). Aggarwal & Chandel (2004) have described how traditional biomass-burning cooking stoves used in the rural areas cause health problems, especially to women and children exposed to smoke emissions.

### *4.7.1 Introducing Improved Cooking Stove*

In Bangladesh, there is a need to develop fuel-efficient wood-burning stoves so as to reduce fuel consumption, improve forest protection and the environment, improve health and hygiene, and reduce the drudgery of women.

Although GHG emission and health hazards from the traditional cooking stoves were not of much concern in the last decades, the growing scarcity of biofuels, which was partly due to their inefficient utilization in traditional cooking stoves, had been the primary motivating factor for the development and dissemination of improved biofuel cooking stoves (Hyman 1994; Rubab & Kandpal 1996). Bhattacharya & Salam (2002) report that new stove designs can improve the efficiency of biomass use for cooking by a factor of 2–3 and save the amount of biomass through increase in biomass use efficiency. It can reduce GHG emissions further through substitution of fossil fuels. Bhattacharya et al. (1999) studied the 'potential of biomass fuel conservation in selected Asian countries' including China, India, Nepal, Pakistan, Philippines, Sri Lanka, and Vietnam. They estimated that through the efficiency improvements in these countries, total biomass fuels saved were 326 m t $yr^{-1}$. In total, the improved ones can save around 152 m t of fuelwood in the domestic sector which is about 43% of the total fuelwood use in the domestic sector. In Bangladesh, a severe dependence on biofuels by a large number of population can potentially diffuse improved cooking stoves, which can save the fuelwood thereby slowing the deforestation and minimizing the GHG emission.

Gowda et al. (1995) showed that if an improved cooking stove replaced the traditional mud-built cooking stove, each family unit could reduce its annual fuelwood consumption by about two-thirds in the rural areas of South India. Rubab & Kandpal (1996) provided some typical numerical calculations revealing that the replacement of the traditional cooking stove by an improved one is financially quite attractive. Smith et al. (1993) described how the Chinese introduced some 129 m improved stoves into rural areas during 1982–1992, mostly biomass cooking stoves, but also coal and space-heating stoves in some areas. Although there were problems with quality control and durability, particularly at the beginning of the program, more than two-thirds of those stoves seemed to be in use during that period. Aggarwal & Chandel (2004) found a positive attitude of the villagers to the concept of improved cooking stoves in India. As the socio-economic condition is almost the same between India and Bangladesh, it can be perceived that the improved cooking stoves can be introduced to the villages of Bangladesh. In addition to this, wood fuel stove project with the improvement of the traditional stoves can well be put on the international 'carbon market' at competitive cost for GHG emission reduction (Ravindranath et al. 2006; Reddy & Balachandra 2006). Since 1978, the Bangladesh

Council of Scientific and Industrial Research (BCSIR) through its Institute of Fuel Research and Development (IFRD) has been trying to develop improved stoves and their accessories in the process of Research and Development (R&D). The IFRD already has developed the improved stoves of three categories, i.e., improved stoves without chimneys, improved stoves with chimneys, and improved stoves with waste heat utilization. Hossain (2003) reports that these improved stoves can save 50–60% fuel and cooking time in comparison to the traditional ones. Through 'advertisement in mass media,' seminars, training programs, and demonstrations, the IFRD is trying to disseminate the improved stove concept throughout the country. But Hossain (2003) states, 'the existing program undertaken by the government, semi-government and non-governmental organizations to disseminate these technologies are not enough to fulfill the demand'.

## 4.8 CDM and Biomass Energy Promotion

Bangladesh has large biomass potential from different sources and is a good candidate for bioenergy technologies. Future bioenergy utilization in developing countries will have to count on modern and efficient technologies, deployed on a commercial basis to guarantee the high quality of energy services needed. In addition, developing countries could become important sources of biomass for other parts of the world. Crop yields are modest in temperate countries when compared with tropical countries, which could become major producers of liquid fuels for world markets. Such a role for developing countries is in line with socio-economic development, global social agenda, and climate change mitigation objectives, while also contributing to increased energy supply security.

Different studies indicate that fossil energy can be beneficially substituted by biomass energy alternatives. One particular advantage of bioenergy in Bangladesh is that it can be organized at small scales, from 1 to 100 MW, thus allowing a slow modular increment in energy supply, avoiding stranded investments, and minimizing risks (Silveira 2005). At a time of restructuring of the electricity sector, these are essential advantages, as economies of scale may not be easily realized in volatile markets. In addition, stakeholder risk aversion and high demand for faster returns will tend to favor smaller projects and a gradual change in the configuration of the electricity infrastructure (Bhattacharya et al. 2003). In addition to this, bioenergy technologies are well tested and have been continuously improved. The projects can be implemented in a decentralized manner, which can be replicable, and tend to enjoy strong support due to the socio-economic and environmental benefits, which can be generated in rural areas.

When fuel substitution and energy efficiency projects are placed under the CDM framework, project financial costs can be reduced and the overall competitiveness of the project tends to be improved. The possibility of generating and trading carbon certificates can help reduce investment risks as these certificates entail hard currency when traded internationally. The CDM may, therefore, help open an investment channel to develop bioenergy projects in Bangladesh, providing an additional

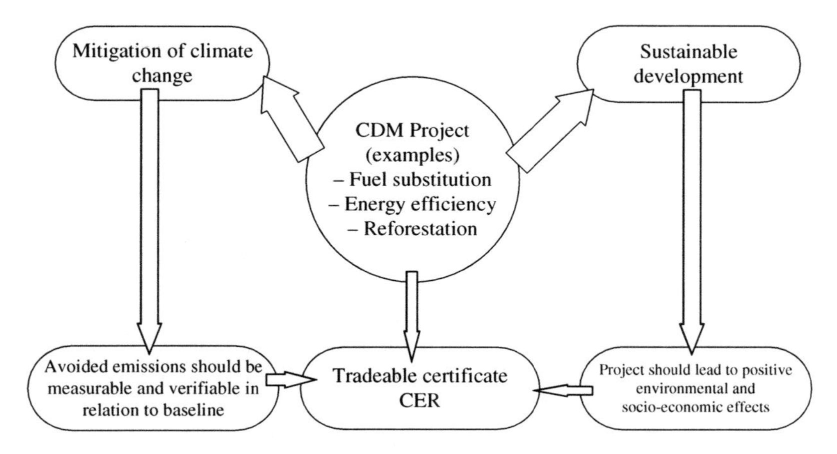


**Fig. 4.6** Bioenergy project in the framework of CDM adopted from Silveira (2005)

tool to foster wider accessibility to modern energy services in these countries and utilizing indigenous energy sources. Environmental management has proved to have a positive effect on the overall management of firms, improving their total performance; the sustainable development dimension of CDM projects may help improve public acceptance to various projects and energy solutions, improving the business environment at regional levels (Fig. 4.6).

As CDM models are further developed, these CDM-related costs should fall, benefiting both project developers and investors as well as enhancing the potential contribution of CDM to climate change mitigation. The CDM-related costs include the business concept, feasibility and baseline studies, approval procedures, and brokerage for marketing carbon credits. Only when carbon markets are better established, replicable CDM models are made available, and transaction costs for developing the CDM project component fall will the additional value of the CDM in the form of tradable credits become a more decisive factor in project finance and implementation. At the early stage, CDM may also serve to integrate components that enhance social and environmental benefits of commercially attractive projects (Ellis et al. 2007; Silveira 2005; Smith & Scherr 2003). Hence, biomass energy is an attractive CDM option because it can easily satisfy both the environmental and the socio-economic requirements imposed on the CDM.

## 4.9 Socio-economic Issues of Biomass Energy Promotion

It has been proved that biomass energy is potential to provide millions of households with income, livelihood activities, and employment. In addition to this, it has a proven role to avoid carbon emission, environment protection, and security of energy supply on a national or regional level. But the primary driving forces include employment or job creation, contribution to regional economy and income improvement (Domac et al. 2005).

Such benefits will result in increased social cohesion and stability that stem from the introduction of an employment and income-generating source. Socio-economic impact of biomass energy promotion in Bangladesh has been shown in Fig. 4.7 adopted from Madlener & Myles (2000).

The promotion of biomass energy in Bangladesh is expected to impact on social aspects, supply, demand, and macro level. All of the activities of biomass promotion will emerge to exert an increased standard of living and social cohesion and stability. With the increased income of the people due to the increased employment opportunity, surrounding environment, their health, and education are easily expected to be developed (Domac et al. 2005). It is evident that rural areas of Bangladesh are suffering from significant levels of outward migration. The bioenergy propensity for rural locations and the deployment of bioenergy plants may have positive effects upon rural labor markets by, first, introducing direct employment and, second, supporting related industries and the employment therein. Finally, it is often possible to achieve significant and sustained development of local initiatives given genuine local involvement of key stakeholders. All the momentum of development will mitigate the rural depopulation achieving sustainable development.

Bioenergy contributes to all important elements of country or regional development: economic growth through business expansion (earnings) and employment;

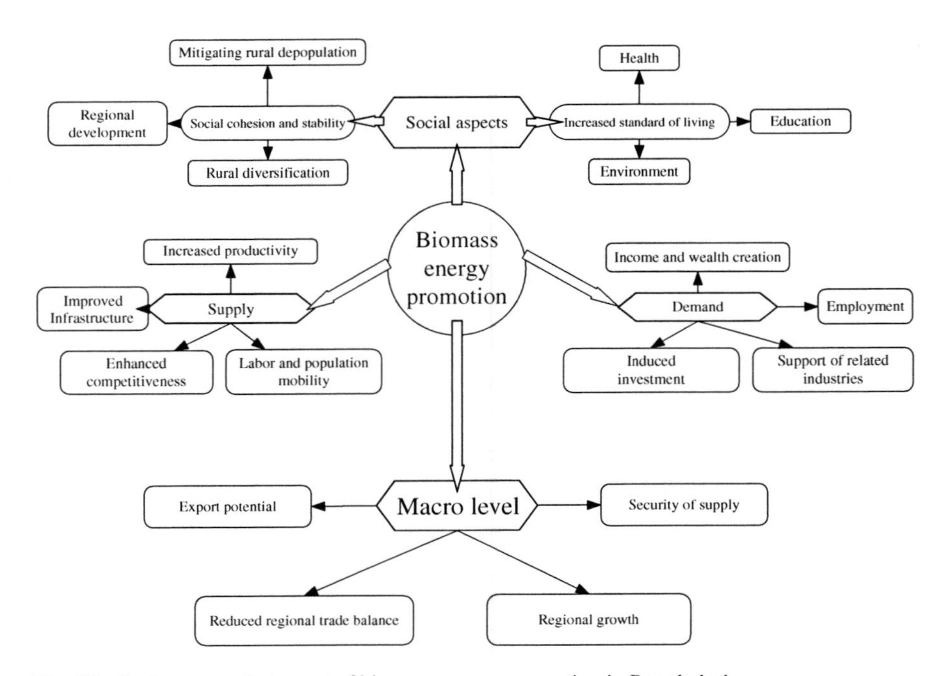


**Fig. 4.7** Socio-economic impact of biomass energy promotion in Bangladesh

import substitution (direct and indirect economic effects on GDP and trade balance); security of energy supply and diversification. In general terms, biomass is better for national and local economies in comparison to fossil fuel, because the fossil fuels are very capital intensive (Domac et al. 2005).

Enhanced productivity, labor, and population mobility are speculated to be the effect of supply, which may result in improved infrastructure in the region. The attractiveness of the investment in the regional small- and large-scale biomass industries, competitiveness, and more efficient entrepreneurship are expected to be developed, which is proven for the enhanced productivity resulting in labor and population mobility in the region. It is expected to emerge income and wealth creation, induced investment, and support to related industries due to the demand side effects. More employment opportunities in the region are expected due to all the activities.

## 4.10 Conclusions

This chapter importantly argues in favor of the 'additionality' of CDM for Bangladesh. It finds that CDM projects can be of fuel substitution and energy efficiency activities in the country. The impact of biomass energy based CDM projects on social cohesion, stability and peoples' standard of living show rationale in favor of the sustainable development in the host country. In production arena of forest biomass, the study proves that Bangladesh has enormous potential to produce biomass in blank, encroached, or other degraded forestlands and homestead compounds. It also shows the potentiality of biomass production technologies. But for the compliance with the CDM projects, these technologies should be harmonized. This chapter indicates that improvement of the traditional cooking stove can save around 50% fuel and cooking time as well as GHG emissions in Bangladesh.

# Chapter 5
# Carbon Sequestration in the Forests of Bangladesh

**Abstract** Bangladesh has huge degraded forestlands which can be reforested by the CDM projects. To realize the potential of the forestry sector in developing countries for full-scale emission mitigation, understanding carbon sequestration potential of different species in different types of plantations are important. This chapter deals with carbon sequestrating potentials of the A/R activities with the possible incorporation with the CDM in Bangladesh. The chapter finds that generally the forests of Bangladesh can sequester 92 tC $ha^{-1}$, on average and 190 $tCha^{-1}$ in the reforested degraded hill forests in particular. This chapter confirms the potential contribution of the carbon sequestration of the forests of Bangladesh, and positively argues the incorporation of forests with the CDM activities. The A/R CDM policy makers in Bangladesh can get this chapter useful to them.

## 5.1 Introduction

Global climate change is one of the most important issues concerning the international community now-a-days. It concerns the increasing accumulation of GHGs, principally $CO_2$, in the atmosphere due to industrial emissions and deforestation (Fearnside 2006; Houghton 2005; Nordell 2003). Forest ecosystems are deemed to be an important factor in climate change because they can be both sources and sinks of atmospheric $CO_2$. They can assimilate $CO_2$ via photosynthesis and store carbon in biomass and in soil (Brown et al. 1996; Graham 2003; IPCC 2000; Lal 2005). Due to the cost-effectiveness in the emission reduction programs having high potential rates of carbon uptake because of the inherent biological growth in the forests and associated environmental and social benefits to the tropical developing countries, increased attention is being focused on tropical forestry to offset carbon emission (IPCC 2000; Moura-Costa 1996; Myers 1996). Tropical forests make up 80% of the total world forests and are recognized as having the greatest long-term potential to sequester atmospheric carbon, via protecting forested lands, slowing deforestation, reforestation, and agroforestry (Brown et al. 1996). Due to deforestation, harvesting, and forest degradation, the world's forests are estimated to have a net source of 1.8 GtC $yr^{-1}$, of which 20% is from tropical deforestation (IPCC 2000). Kram

Md.D. Miah et al., *Forests to Climate Change Mitigation*, Environmental Science and Engineering, DOI 10.1007/978-3-642-13253-7_5, © Springer-Verlag Berlin Heidelberg 2011 

et al. (2000) point out that the distribution of both income and GHG emissions is very unbalanced between various world regions. The relative importance of individual gases and sources of emissions differs from region to region. Kram et al. (2000) analyzed that currently developing countries account for about 46% of all emissions, but by 2100 no less they would contribute 67–76% of the global total, while the total income generated in these countries would reach 58–71% from only 16% in 1990. But Kram et al. (2000) conclude that when population size and the levels of affluence in the developing countries confront with the potential severity of climate change-induced damages, the scenarios are very different. Higher population densities and lower income make the countries more vulnerable for adverse climate change impacts, and that lower income creates less favorable conditions for mitigation and/or adaptation measures. Domestic activity in developed countries must, therefore, receive priority, and the main emitters of $CO_2$ should assume their responsibility for tackling the causes. In the Protocol, which was adopted in Kyoto, Japan, in 1997, industrialized countries have committed themselves to reduce their combined GHG emissions by more than 5% relative to the level in 1990, in the period between 2008 and 2012.

The Kyoto Protocol recognizes forestry and land-use change activities as sinks and sources of atmospheric carbon. In a special report on LULUCF, the IPCC concludes that activities in the realm of land-use change and forestry provide an opportunity to affect the carbon cycle positively. First, they may contribute to reducing GHG emissions by avoiding deforestation or improving forest management. Second, forestry and land-use change may increase the uptake of atmospheric carbon (commonly referred to as carbon sequestration) through afforestation and reforestation (Graham 2003). Under the agreement reached at the COP9, industrialized countries will be able to meet a part of their emission reduction commitments under the Kyoto Protocol by financing reforestation and afforestation activities (AR) in developing countries through the CDM (UNFCCC 2006b). The Kyoto Protocol states that CDM projects should assist developing countries in achieving sustainable development.

In 2000, the total plantation area of Bangladesh under reforestation activities was 625,000 ha, with an annual planting rate of 22,000 ha (FAO 2001). With a huge pool of existing plantation and natural forests in Bangladesh, it can be assumed that Bangladesh is playing a major role in mitigating global warming. Bangladesh has a lot of degraded forestlands and other wastelands to be reforested, but severe poverty and lack of appropriate technology in the country are the barriers for establishing carbon sinks in the forests through A/R.

Quantification of net carbon sequestration by plantation species is a prime research question in deducing the carbon credit of reforested plantations. Various authorities have initiated small-scale plantation schemes during the 1970s with the aim of reforesting the denuded and semi-denuded hills of the Chittagong areas using economically important and ecologically suitable tree species (pers. commun.). The plantations are composed of various indigenous and fast-growing exotic species and, as many of them were established successfully, the area is now viewed as an important example for carbon sequestration in Bangladesh.

This chapter was based on field data collection through physical measurement, field observation, and laboratory analysis. The field study was conducted between February 2001 and June 2002. The situation in Bangladesh regarding the trend in global warming was also analyzed. The implications of carbon trading for Bangladesh are discussed – highlighting the problems, constraints, and uncertainty. Finally, recommendations have been made to expedite the participation of Bangladesh in global carbon trading. In addition, the study investigates the general status of the carbon stock in the whole of Bangladesh, based on the biomass stock with a special attention to Chittagong hilly region. There is some degree of uncertainty and lack of policy regarding Bangladesh's participation in carbon trading under the CDM. This is also discussed and recommendations made for the implementation of proper policy decisions.

## 5.2 Materials and Methods

To determine the general situation on carbon sequestration by the forests of Bangladesh, information on dry matter (dm) quantity has been collected from secondary sources (ESSD 1998) and calculated for carbon stocks. To demonstrate carbon sequestration, the Chittagong region (Fig. 4.2) was deliberately selected because this region had the largest areas of degraded forests which can be reforested through CDM projects. To quantify the carbon stock within the Chittagong region, plantations of 13 tree species, with ages ranging from 6 to 23 years, managed by Chittagong University, were studied. The elevation of the study areas ranged between 14 and 87 m above mean sea level. The area was once covered with dense tropical semi-evergreen forests of *Dipterocarpus* spp., *Artocarpus* spp., and *Albizia* spp., which had been deforested by overexploitation and encroachment (Islam et al. 1999). Aggressive grasses and shrubs, such as *Imperata cylindrica*, *Saccharum spontaneum*, *Lantana camara*, and *Melastoma* spp., have replaced the forest vegetation. The majority of planted species were indigenous and only a few were exotic. The climate is tropical monsoon with more than 10°C difference between summer and winter temperatures. Mean annual rainfall is about 280 cm, which is received almost entirely in March to June. Soils of the study site belonged to brown hill soil (yellowish red, coarse, isohyperthermic Typic Dystrochrept), which are developed in tertiary hill sediments of unconsolidated and partially consolidated beds of sandstones, siltstones, and conglomerates (Gafur et al. 1979; SSS 1979). Soils are shallow and sandy loam in texture at the surface, followed by compacted ferruginous layers at 1.5–2.7 m depths (SSS 1979). In some places, compacted layers have been exposed due to accelerated erosion after deforestation. Accelerated erosion has accounted for a substantial loss of silt and clay particles from upland areas, which in turn are deposited in the valleys. Induced leaching and surface runoff of basic nutrients (Ca, Mg, and K) have caused upland surface soils to gradually increase in acidic cations (Fe and Al), and, as a result, degraded soils have become distinctly acidic over time (Gafur et al. 1979). Degraded soils have higher bulk densities and are deficient in organic matter and total N. It contains higher amounts of ferruginous concretions compared to soils under well-stocked natural forests (Islam et al. 1999). For reforestation purpose, seeds were germinated in $15 \times 10$ cm$^2$ plastic bags filled with a mixture of sand, forest topsoil, compost, and ash. All seedlings were grown in the nursery managed by the forestry research wing of Chittagong University. Four-month-old seedlings were transplanted in pure stands at $1.82 \times 1.82$ m$^2$ spacing. Weeding continued once per year through the second year. No fertilizer was applied in the plantation.

### *5.2.1 Sampling Procedures*

The study covered 13 plantation species, comprising six exotic and seven indigenous species. As all of the plantations were in the hills, sampling of the plots was done in all slopes of the hills, i.e., bottom hill, middle hill, and top hill. From each stand, a total of nine plots were sampled taking every three from bottom, middle, and top slope. The size of the sample plot was $10 \times 10$ m$^2$. A total of 252 sample plots were studied from the 28 stands of different years established in 45.44 ha of lands. The plantation species were selected based on the higher frequency of the species used in the reforestation throughout the Chittagong hilly region. The data on the frequency of reforestation were collected from the research division of the Forestry Department of Bangladesh (pers. commun.). The age of the sampled stands was not uniform for all the species, because different stands were established in different times in the study area forming the different stands with different ages (Table 5.1). The selection of the different aged stands was based on the availability in the study area. Among them *Acacia auriculiformis* was selected for the seven stands ranging from 6 to 18 years; *Dipterocarpus turbinatus* for four ranging from 6 to 23 years; *Swietenia mahagoni* for three ranging from 11 to 15 years; *Aphanamixis polystachya*, *Chukrasia tabularis*, *Eucalyptus camaldulensis*, and *Lagerstroemia speciosa* were for two for each species; and *Acacia mangium*, *Albizia procera*, *Gmelina arborea*, *Pinus caribaea*, *Syzygium grande*, and *Tectona grandis* were for one for each. The average stand density was 1267 ha$^{-1}$ ranging from 171 to 2594 ha$^{-1}$ (Table 5.1). The reasons for the understocking of some stands might be for overextraction of forest resources, encroachment, and natural calamities, like cyclone and tornado. However, the reasons for the understocking of the stands were not studied.

### *5.2.2 Procedure of Net Carbon Sequestration Estimation*

In a plantation, gross carbon content includes the carbon content in biomass, soil, and fallen litters. In the study of the 13 plantation tree species, annual growth of the biomass (tree biomass and litter biomass) was measured to get the annual carbon increment. Annual litter and fuelwood collection was considered as the prime cause

**Table 5.1** Stand ages, densities, and plot number sampled in the Chittagong hilly region, Bangladesh

| Species | Age of the stand | Stand density (tree ha$^{-1}$) | Average of the stand density (tree ha$^{-1}$) | Plot number sampled ($n$) | | |
|---|---|---|---|---|---|---|
| | | | | Tree biomass (10 × 10 m) | Fallen litter biomass (2 × 2 m) | Soil carbon (1 × 1 m) |
| *A. auriculiformis* | 6 | 971 (76) | 970 (39) | 7 × 3 × 3 = 63 | 3 | 7 × 3 = 21 |
| | 8 | 1259 (57) | | | 2 | |
| | 11 | 1051 (64) | | | 1 | |
| | 15 | 895 (23) | | | 1 | |
| | 16 | 1038 (115) | | | 3 | |
| | 17 | 839 (112) | | | 2 | |
| | 18 | 736 (154) | | | 1 | |
| *A. mangium* | 11 | 1103 (221) | 1103 (221) | 3 × 3 = 9 | 1 | 3 |
| *A. procera* | 20 | 468 (81) | 468 (81) | 3 × 3 = 9 | 1 | 3 |
| *A. polystachya* | 12 | 2028 (64) | 1933 (90) | 2 × 3 × 3 = 18 | 1 | 2 × 3 = 6 |
| | 13 | 1838 (167) | | | 1 | |
| *C. tabularis* | 8 | 1512 (189) | 1288 (118) | 2 × 3 × 3 = 18 | 1 | 2 × 3 = 6 |
| | 17 | 1063 (106) | | | 1 | |
| *D. turbinatus* | 6 | 809 (67) | 1044 (83) | 4 × 3 × 3 = 36 | 3 | 4 × 3 = 12 |
| | 8 | 1679 (130) | | | 2 | |
| | 13 | 797 (134) | | | 1 | |
| | 23 | 892 (115) | | | 1 | |
| *E. camaldulensis* | 8 | 1370 (78) | 1413 (129) | 2 × 3 × 3 = 18 | 1 | 2 × 3 = 6 |
| | 18 | 1455 (253) | | | 1 | |
| *G. arborea* | 13 | 1826 (219) | 1826 (219) | 3 × 3 = 9 | 1 | 3 |
| *L. speciosa* | 18 | 1834 (102) | 1646 (106) | 2 × 3 × 3 = 18 | 1 | 2 × 3 = 6 |
| | 20 | 1457 (170) | | | 1 | |
| *P. caribaea* | 18 | 1279 (221) | 1279 (221) | 3 × 3 = 9 | 1 | 3 |
| *S. mahagoni* | 11 | 2091 (58) | 1629 (110) | 3 × 3 × 3 = 27 | 1 | 3 × 3 = 9 |
| | 12 | 1239 (116) | | | 1 | |
| | 15 | 1559 (234) | | | 1 | |
| *S. grande* | 13 | 1744 (92) | 1744 (92) | 3 × 3 = 9 | 1 | 3 |
| *T. grandis* | 20 | 651 (158) | 651 (158) | 3 × 3 = 9 | 3 | 3 |
| | Total number of stand = 28 | Average = 1267 (37) | | $N = 252$ | $N = 39$ | $N = 84$ |

Figures in parentheses indicate the SE

of carbon loss from the plantations. To get the tonne carbon per year (tC $yr^{-1}$), the loss, through litter and fuelwood collection, was deducted from the gross carbon content.

### 5.2.3 *Biomass Estimation of the Plantations*

The aboveground biomass of trees in general in the stands, for every plot, was measured by the following formula (Brown et al. 1989):

$$Y = \exp\left\{-2.4090 + 0.9522\ln(D^2HS)\right\}$$

where

exp [. . .] means 'raised to the power of [. . .]'
$Y$ = aboveground biomass in kg
$H$ = height of the trees in m
$D$ = Diameter at breast height (1.3 m) in cm
$S$ = Oven dry density in units of t $m^{-3}$ for a specific species (Brown 1997; Sattar et al. 1999)

Underground biomass was calculated as 15% of the aboveground biomass (IPCC 2003). The above- and underground biomasses were added to get the total biomass of the stand.

### 5.2.4 *Procedure of the Biomass Estimation of the Litterfall*

To estimate the biomass of fallen litter in all stands, sample plots of 2×2 $m^2$ were established in the crown-covered area. A total of 39 plots were established and studied for the 28 stands (Table 5.1). In the study, it was assumed that there was a negligible difference of litterfall quantity due to the slope positions in the stands. Most of the plots were established at the mid-hill position of each stand. However, based on the ease of establishing the permanent plot, some plots were also established in bottom and top slopes.

All weeds and brush from the plots were cleared and the debris burnt. Fallen litter was collected at intervals of 15 days. All the litter of a month was gathered for each plot. The litter included leaves, fruits, seeds, barks, and twigs. The fresh weight of the collected litter was measured in the field using a field balance. A fresh sample from each plot was oven-dried at 65°C until a constant weight was reached. The fresh weight was converted to dry weight and average litterfall of 6 months was converted to annual litterfall per hectare to estimate the biomass of the litterfall in the stands.

### 5.2.5 Carbon Content Estimation

Carbon content of tree tissue and fallen litter was measured based on the biomass of trees, with the following formula (IPCC 2003):

$$\text{Biomass carbon content (t ha}^{-1}\text{)} = \text{Biomass weight (tdm ha}-1\text{)} \times 0.5\text{tC tdm}-1$$

where tC = tonne carbon; tdm = tonne dry matter

### 5.2.6 Procedure of the Estimation of Soil Carbon Content

To estimate the carbon content in the soil of the plantations, soil samples and humus were collected from all the studied plantations. From each stand, a total of three soil plots were sampled taking one from bottom, one from middle, and one from top slopes. The size of the sample plot was $1 \times 1$ $m^2$. Each sample was a composite of three subsamples. Thus for the 28 stands, a total of 84 soil samples were studied (Table 5.1). Soil samples were collected at a depth of 30 cm using an earth augur. The samples were carefully taken to the laboratory for chemical analysis. To estimate the percentage of organic carbon in the soil and humus, samples were analyzed through the wet oxidation method (Petersen 1996).

Then, the soil organic carbon was converted to tC $ha^{-1}$ using the following formula:

$$\text{Organic carbon per hectare} = 40 \times O_{c\%} \text{ t ha}^{-1}\text{(30 cm depth)}$$

Here, $O_{c\%}$ = organic carbon percentage (Donahue et al. 1987).

### 5.2.7 Procedure for the Estimation of Carbon Loss Through Forest Product Extraction from the Stands

First, a reconnaissance survey was conducted to determine the types of forest product extracted by people from the plantations within the study area. A total of five points were carefully selected which were considered as the major exit points from the forests. Five working groups, consisting three members each, were appointed at every point. Data were collected 1 day per week from 07:00 to 19:00 hours. The day of each week was selected randomly. The types of collected litter and fuelwood were determined by personal observation, as well as with the help of the collectors. The fresh weight of the collected litter and fuelwood was recorded by species separately. The species that were under study in the stands for biomass and litterfall were only considered. The calculation of the dry matter lost was calculated for species. In this case, stands and slopes were not considered. But it was confirmed that all fuelwood and litters were extracted from at least one of our sampled stands. To convert fresh weight to dry weight, 1% of both litter and fuelwood of each species was oven-dried at 65°C in the laboratory until a constant weight was obtained. Thus, the total fresh weight of the litter and fuelwood was converted to total dry weight of each type. The total dry weight of all types, for all the points, gave the total biomass loss per day. The biomass of a day was converted to the yearly biomass loss from the studied plantations.

### 5.2.8 Statistical Analysis

To show the difference of total carbon contents among the stands and plantation species separately, one-way analysis of variance (ANOVA) was carried out. Linear regression analysis was also carried out to show the effect of soil and litter carbon contents and age on the total carbon contents. Correlation test was carried out to show the relationship between the variables. All the analysis was carried out using SPSS 13.0 for Windows.

## 5.3 Results and Discussion

### 5.3.1 Potential of Bangladeshi Forests in Carbon Uptake

Bangladesh has a number of diversified ecosystems, viz., wetlands, rain forests, moist deciduous forests, semi-arid areas, and mangroves, each of which contains special types of plants and acts as the important carbon sink. Floral diversity is richer in the hill forests of Chittagong, Chittagong Hill Tracts, and Sylhet. Wetlands comprise about half the area of Bangladesh, forming a unique mosaic of habitats, with a rich diversity of flora, some of which are, as yet, unexplored. The forestlands of Bangladesh are broadly categorized as government forestland and private forestland, covering 2.2 and 0.4 m ha of lands, respectively. Of the government forestlands, 1.3 m ha of natural forests and plantations are under the jurisdiction of the Forest Department within the Ministry of Forests and Environment. The Forest Department of Bangladesh manages three types of forests, i.e., (a) tropical evergreen or semi-evergreen forests (640,000 ha) in the eastern district of Chittagong, Cox's Bazar, Sylhet, and the Chittagong Hill Tracts; (b) moist or dry deciduous forests, known as Sal forest (*Shorea robusta*), situated mainly in the central parts and freshwater areas in the northeastern region; and (c) tidal mangrove forests along the coastal zone (520,000 ha) to the southwest of Khulna and other mangroves along the Chittagong and Noakhali coastal belts (Gain 1998).

It is estimated that more than 5,000 species of higher plants occur in Bangladesh. Thick foliage and species diversity have made Bangladesh one of the richest flora regions in the world. It is estimated that around 300 plant species are found in the wetlands of Bangladesh (SEHD 1998). It is reported that the homestead flora of Bangladesh provide about 70% of all wood consumed and 90% of all fuelwood and bamboo (Alam et al. 1990). It is also estimated that around 149 village tree species exist in Bangladesh (Khan & Alam 1996).

**Table 5.2** Biomass and carbon density in the forests of Bangladesh

| Forest types | Aboveground biomass (tdm $ha^{-1}$) | Underground biomass (15% of the aboveground) (tdm $ha^{-1}$) | Total biomass (tdm $ha^{-1}$) | Carbon stock (tC $ha^{-1}$) (50% of the dry matter) |
|---|---|---|---|---|
| Closed large-crowns | 206–210 | 32 | 242 | 121 |
| Closed small-crowns | 150 | 23 | 173 | 87 |
| Disturbed closed | 190 | 29 | 219 | 110 |
| Disturbed open | 85 | 13 | 98 | 49 |
| Average | | | | 92 |

The present study estimates that, on average, 92 tC $ha^{-1}$ (Table 5.2) is stored by the existing tree tissue in the forests of Bangladesh. The specific figures are as follows: closed large-crown forests 121 tC $ha^{-1}$, closed small-crown forests 87 tC $ha^{-1}$, disturbed closed forests 110 tC $ha^{-1}$, and disturbed open forests 49 tC $ha^{-1}$. Forest soils in Bangladesh, as well as Asia, store at a rate of 115,100 and 60 tC $ha^{-1}$ in moist and seasonal and dry soils, respectively (ESSD 1998). But these figures are expected to be reduced very promptly due to the unfavorable human dimensions on forests like overextraction and encroachment. The degradation of the forests seems to be uncontrolled, thereby large tracts of forestlands are expected to be vacant in the near future. But due to the lack of funds and technology, the government of Bangladesh is not capable of reforesting the degraded forestlands. Only CDM projects can help reforest the degraded forestlands and achieve the sustainable development in the country through the revenues earned by selling the carbon credits. Even though reducing emissions from deforestation and degradation is not creditable in the first commitment period (2008–2012) of the Kyoto Protocol, it is expected to be included in the second commitment period (2012+). The main focus of this study is to show the potentialities of reforestation expected to be implemented by the CDM projects in Bangladesh in the first commitment period.

### *5.3.2 Gross Carbon Content in the Stands*

The average highest biomass carbon content (145 tC $ha^{-1}$, Standard error of mean (SE) 7.73) was found in the *A. polystachya* stands, and the lowest (43 tC $ha^{-1}$, SE 7.70) was found in the *S. mahagoni* stands. It was found that 8-year-old *A. auriculiformis* stand had the highest (173 tC $ha^{-1}$, SE 64.82) biomass carbon contents followed by 12-year-old *A. polystachya* stand (166 tC $ha^{-1}$, SE 6.86). The lowest (10 tC $ha^{-1}$, SE 2.31) carbon contents were found in the 6-year-old *D. turbinatus* stand. The average highest soil (including humus) carbon content (113 tC $ha^{-1}$, SE 5.39) was found in the *L. speciosa* stands, while the lowest was found in the *P. caribaea* plantation (83 tC $ha^{-1}$, SE 6.93). The highest soil carbon content (132 tC $ha^{-1}$, SE 6.81) was found in the 23-year-old *D. turbinatus* stand followed by 20-year-old *L. speciosa* stand (120 tC $ha^{-1}$, SE 7.13). The lowest soil carbon content (82 tC $ha^{-1}$) was found in the 8-year-old *C. tabularis*, 6-year-old *D. turbinatus*, and 8-year-old *E. camaldulensis* stands with the SE ranging from 3.53 to 7.75. All the species had average litter carbon contents as 3 tC $ha^{-1}$, except *S. mahagoni*, *P. caribaea*, and *S. grande* stands, which had only 2 tC $ha^{-1}$ carbon with SE ranging from 0 to 0.03. The stands of 17- and 18-year-old *A. auriculiformis* and 13-year-old *A. polystachya* had the highest fallen litter carbon contents (4 tC $ha^{-1}$) with the SE ranging from 0 to 0.22; the lowest (2 tC $ha^{-1}$) was found in the 6-, 11-, and 15-year-old *A. auriculiformis*, 23-year-old *D. turbinatus*, 18-year-old *P. caribaea*, 11-, 12-, and 15-year-old *S. mahagoni*, and 13-year-old *S. grande* stands with the SE ranging from 0.38 to 1.8.

The study also showed that the highest total carbon content (255 tC $ha^{-1}$, SE 7.21) was found in the *A. polystachya* plantations, followed by concentrations of 247 (SE 31.96), 224 (SE 25.76), 211 (SE 22.77) and 203 (SE 31.45) tC $ha^{-1}$ for the *L. speciosa*, *T. grandis*, *A. procera*, and *E. camaldulensis* plantations, respectively. The lowest carbon content (140 tC $ha^{-1}$, SE 12.42) was found in the *S. grande* plantation (Table 5.3). The average carbon content for all the stands studied was 190 (SE 5.72) tC $ha^{-1}$. Considering the stands of different ages and species, it was shown that 12-year-old *A. polystachya* had the highest total carbon contents (269 tC $ha^{-1}$, SE 6.12) followed by 8-year-old *A. auriculiformis* (268 tC $ha^{-1}$, SE 66.49), while the lowest (96 tC $ha^{-1}$, SE 5.74) was observed in the 6-year-old *D. turbinatus* stand (Table 5.3).

One-way ANOVA showed that total carbon contents in the stands significantly varied for both $p < 0.05$ and $p < 0.01$, for both plantation species and stands with different ages. Among the species, *A. polystachya and L. speciosa* had been shown to have the total carbon contents significantly ($p < 0.05$) higher than the other mentioned species (Table 5.3). Among the stands of the same species, 8-year-old *A. auriculiformis* in the bottom and middle hill and 13-year-old *D. turbinatus* in the middle hill were significantly ($p < 0.05$) higher than the other mentioned species (Table 5.3). ANOVA also showed that soil carbon contents for both the species and stands were significantly ($p < 0.05$) different for the three different hill positions, i.e., bottom, middle, and top. It was found that among the slopes, bottom hill stocked more carbon than middle and top slopes.

The impact of hill position has been shown in Figs. 5.1 and 5.2. The correlation analysis showed that soil carbon and total carbon contents were significantly ($p < 0.05$ and $p < 0.01$) correlated ($r = 0.685$). Age of the stands and total carbon contents were significantly ($p < 0.05$ and $p < 0.01$) correlated ($r = 0.386$). Fallen litter carbon content was also significantly ($p < 0.05$) correlated with total carbon contents. The analysis also showed that soil carbon and fallen litter carbon contents were significantly ($p < 0.05$) correlated. The relationship matrix of the mentioned variables has been shown in Fig. 5.3. Linear regression analysis showed that soil carbon contents had a highly significant ($p < 0.05$, $p < 0.01$, and $p < 0.001$) effect (adjusted $r^2 = 0.434$) on the total carbon contents.

A study, conducted on the biomass and nutrient distribution in *A. auriculiformis*, *D. turbinatus*, and *P. caribaea* stands on the Chittagong University campus, showed that the total aboveground biomass for 4-year-old *A. auriculiformis*, 5-year-old *A. auriculiformis*, 8-year-old *D. turbinatus*, and 4-year-old *P. caribaea* was 76, 51, 32, and 62 t ha$^{-1}$, respectively (Osman et al. 1992). They also found the organic carbon varies among the three hill positions. They reported that organic carbon was highest in the bottom followed by middle and top slopes. It might be due to the nutrients leaching from the top to the bottom through middle slope. Miah et al. (2001) and Gafur et al. (1979) found the same trend of nutrient distribution in the slopes of the plantation. Their reports espouse our results also. In the present study, average organic carbon percentage was 1.04 for all the stands.

Osman et al. (2001) conclude that soil organic matter increases with the age of the plantation, until canopy closure, but is dependent on the ability of the species to produce litter. Another study showed that organic carbon, stored in some Indian soils, was 41, 120, 13, and 18 tC ha$^{-1}$ for red, laterite, saline, and black soils, respectively (Jha et al. 2001).

That study also reported that soil organic carbon content, under different land-use regimes in India, was found to be 120, 40, and 40 tC ha$^{-1}$ for forest, agriculture, and pasture, respectively (Jha et al. 2001). The variation of the total biomass, soil, and total carbon contents per hectare was mainly influenced by the density of trees per hectare. However, the causes of variation of the density of trees per hectare for the species and the different aged stands were not studied.

### *5.3.3 Annual Increment in Carbon Contents in the Plantations*

The study found a net 4 (SE 0.31) tC ha$^{-1}$yr$^{-1}$ increment in the plantations, considering productivity and loss of litter and fuelwood from the plantation. The highest net mean annual increment (MAI) in carbon stock was found to be 9.83 (SE 1.50) tC ha$^{-1}$yr$^{-1}$ in the *E. camaldulensis* plantation, followed by *A. mangium* (7.48 tC ha$^{-1}$yr$^{-1}$, SE 0.66), while the lowest was in the *G. arborea* plantation, at 0.25 tC ha$^{-1}$yr$^{-1}$ (SE 0.64) (Table 5.4). Considering the different aged stands it was shown that 8-year-old *A. auriculiformis* and 8-year-old *E. camaldulensis* had the highest (10 tC ha$^{-1}$yr$^{-1}$, SE ranging from 1.30 to 3.06) net increment followed by 18-year-old *E. camaldulensis*; while the lowest was in the *G. arborea* plantation as described earlier (Table 5.4). *A. polystachya* (13 years old) and *L. speciosa* (20 years old) were found to have the negative value of net MAI of carbon stock in the stands. It is clear from Table 5.4 that fuelwood and litter collection of these two species are comparatively higher than the other species. Carbon loss was found to vary greatly among the different plantations, due to the large species diversity in the areas concerned. The large volumes of wood and litter collection of *A. polystachya*, *L. speciosa*, and *G. arborea* was most probably due to the high burning value of those species (Ara et al. 1989; Bhuiyan et al. 1995).

**Table 5.3** Gross carbon stocks in the stands in the Chittagong hilly region, Bangladesh

| Stand of the trees | | Aboveground biomass (t ha$^{-1}$) | | | Underground biomass (t ha$^{-1}$) | | | Total above and underground biomass (t ha$^{-1}$) | | |
|---|---|---|---|---|---|---|---|---|---|---|
| Plantation species | Age | Bottom | Middle | Top | Bottom | Middle | Top | Bottom | Middle | Top |
| *A. auriculiformis*[a] | 06[a1] | 53.40 | 47.22 | 15.66 | 8.01 | 7.08 | 2.35 | 61.41 | 54.31 | 18.01 |
| | | (0.41) | (11.84) | (4.95) | (0.06) | (1.78) | (0.74) | (0.47) | (13.62) | (5.69) |
| | 08[a2] | 677.64 | 219.12 | 6.69 | 101.65 | 32.87 | 1.00 | 779.29 | 251.99 | 7.70 |
| | | (183.60) | (34.79) | (0.64) | (27.54) | (5.22) | (0.10) | (211.14) | (40.01) | (0.73) |
| | 11[a3] | 181.04 | 118.34 | 41.44 | 27.16 | 17.75 | 6.22 | 208.20 | 136.09 | 47.65 |
| | | (16.67) | (9.99) | (3.67) | (2.50) | (1.50) | (0.55) | (19.17) | (11.49) | (4.22) |
| | 15[a4] | 376.20 | 139.09 | 16.12 | 56.43 | 20.86 | 2.42 | 432.64 | 159.95 | 18.54 |
| | | (40.05) | (18.87) | (2.40) | (6.01) | (2.83) | (0.36) | (46.06) | (21.70) | (2.76) |
| | 16[a5] | 366.62 | 170.90 | 33.42 | 54.99 | 25.64 | 5.01 | 421.62 | 196.54 | 38.43 |
| | | (17.04) | (16.35) | (9.09) | (2.56) | (2.45) | (1.36) | (19.60) | (18.80) | (10.45) |
| | 17[a6] | 245.68 | 184.81 | 11.22 | 36.85 | 27.72 | 1.68 | 282.54 | 212.54 | 12.90 |
| | | (40.16) | (13.95) | (1.07) | (6.02) | (2.09) | (0.16) | (46.18) | (16.05) | (1.23) |
| | 18[a7] | 186.25 | 131.70 | 53.45 | 27.94 | 19.76 | 8.02 | 214.19 | 151.46 | 61.47 |
| | | (39.04) | (17.96) | (29.48) | (5.86) | (2.69) | (4.42) | (44.89) | (20.66) | (33.90) |
| *A. mangium*[b] | 11 | 196.67 | 197.67 | 103.66 | 29.50 | 29.65 | 15.55 | 226.17 | 227.32 | 119.21 |
| | | (6.36) | (16.22) | (14.30) | (0.95) | (2.43) | (2.15) | (7.31) | (18.65) | (16.45) |
| *A. procera*[c] | 20 | 309.30 | 96.38 | 122.33 | 46.39 | 14.46 | 18.35 | 355.69 | 110.84 | 140.68 |
| | | (34.49) | (14.03) | (16.76) | (5.17) | (2.10) | (2.51) | (39.66) | (16.13) | (19.27) |
| *A. polystachya*[d] | 12[d1] | 285.70 | 251.89 | 326.94 | 42.86 | 37.78 | 49.04 | 328.56 | 289.67 | 375.98 |
| | | (5.78) | (12.30) | (10.42) | (0.87) | (1.84) | (1.56) | (6.65) | (14.14) | (11.99) |
| | 13[d2] | 251.34 | 169.31 | 226.79 | 37.70 | 25.40 | 34.02 | 289.04 | 194.71 | 260.81 |
| | | (35.05) | (11.95) | (20.77) | (5.26) | (1.79) | (3.12) | (40.31) | (13.74) | (23.89) |
| *C. tabularis*[e] | 08[e1] | 244.66 | 46.87 | 0.17 | 36.70 | 7.03 | 0.03 | 281.36 | 53.90 | 0.2 |
| | | (42.59) | (20.94) | (0.02) | (6.39) | (3.14) | | (48.98) | (24.08) | (0.02) |
| | 17[e2] | 303.53 | 177.94 | 6.82 | 45.53 | 26.69 | 1.02 | 349.06 | 204.63 | 7.84 |
| | | (2.79) | (17.01) | (1.62) | (0.42) | (2.55) | (0.24) | (3.21) | (19.56) | (1.86) |
| *D. turbinatus*[f] | 06[f1] | 27.99 | 23.06 | 2.86 | 4.20 | 3.46 | 0.43 | 32.19 | 26.52 | 3.29 |
| | | (3.69) | (1.30) | (0.61) | (0.55) | (0.19) | (0.09) | (4.24) | (1.49) | (0.70) |
| | 08[f2] | 149.10 | 82.02 | 2.98 | 22.37 | 12.30 | 0.45 | 171.47 | 94.32 | 3.43 |
| | | (22.24) | (23.83) | (0.48) | (3.34) | (3.57) | (0.07) | (25.58) | (27.40) | (0.55) |

| Total carbon contents of above and underground biomass (tC ha$^{-1}$) | | | Soil carbon contents (tC ha$^{-1}$) | | | Fallen litter carbon contents (tC ha$^{-1}$yr$^{-1}$) | | | Total carbon contents (tC ha$^{-1}$) | | | |
|---|---|---|---|---|---|---|---|---|---|---|---|---|
| Bottom | Middle | Top | Bottom | Middle | Top | Bottom | Middle | Top | Bottom mean | Middle | Top | Species |
| 30.71 (0.23) | 27.15 (6.81) | 9.00 (2.85) | 89 | 83 | 77 | 2.43 (0.75) | 2.43 (0.75) | 2.43 (0.75) | 122.14 (0.96) | 111.79 (6.71) | 88.44 (3.60) | 189.56 (13.03) |
| 389.64 (105.57) | 126.00 (20.00) | 3.85 (0.37) | 97 | 94 | 83 | 3.27 (0.23) | 3.03 (0.23) | 3.03 (0.23) | 489.91[1] (105.59) | 223.03[a1] (19.96) | 89.88 (0.48) | |
| 104.10 (9.59) | 68.04 (5.74) | 23.83 (2.11) | 104 | 96 | 86 | 2.30 | 2.30 | 2.30 | 210.40 (9.59) | 166.34 (5.74) | 112.13 (2.11) | |
| 216.32 (23.03) | 79.98 (10.85) | 9.27 (1.38) | 104 | 95 | 87 | 2.30 | 2.30 | 2.30 | 322.62 (23.03) | 177.28 (10.85) | 98.57 (1.38) | |
| 210.81 (9.80) | 98.27 (9.40) | 19.22 (5.23) | 107 | 100 | 94 | 3.50 | 2.90 (0.60) | 2.97 (0.63) | 321.31 (9.80) | 201.17 (9.27) | 116.18 (5.86) | |
| 141.27 (23.09) | 106.27 (8.02) | 6.45 (0.62) | 110 | 104 | 102 | 3.20 | 4.50 | 4.50 | 254.47 (23.09) | 214.77 (8.02) | 112.95 (0.62) | |
| 107.09 (22.45) | 75.73 (10.33) | 30.74 (16.95) | 112 | 107 | 102 | 4.30 | 4.30 | 4.30 | 223.39 (22.45) | 187.03 (10.33) | 137.04 (16.95) | |
| 113.08 (3.66) | 113.66 (9.33) | 59.60 (8.22) | 107 | 96 | 83 | 3.00 | 3.00 | 3.00 | 222.08 (3.66) | 211.95 (9.31) | 146.60 (8.55) | 193.55 (12.42) |
| 177.85 (19.83) | 55.42 (8.07) | 70.34 (9.63) | 117 | 116 | 88 | 3.00 | 3.00 | 3.00 | 297.85 (19.83) | 174.42 (8.07) | 161.34 (9.63) | 211.20 (22.77) |
| 164.28 (3.33) | 144.83 (7.07) | 187.99 (5.99) | 118 | 101 | 82 | 2.83 (0.17) | 2.50 | 3.50 | 285.11 (3.19) | 248.33 (7.07) | 273.49 (5.99) | 255.29[2] |
| 144.52 (20.15) | 97.36 (6.87) | 130.40 (11.94) | 129 | 113 | 100 | 3.50 | 3.50 | 3.50 | 277.02 (20.15) | 213.86 (6.87) | 233.90 (11.94) | (7.21) |
| 140.68 (24.49) | 26.95 (12.04) | 0.10 (0.01) | 89 | 84 | 77 | 3.30 | 3.30 | 3.30 | 230.98 (24.49) | 113.25 (12.04) | 80.05 (1.18) | 169.69 (19.16) |
| 174.53 (1.61) | 102.32 (9.78) | 3.92 (0.93) | 111 | 105 | 90 | 2.70 | 2.70 | 2.70 | 287.23 (1.61) | 210.02 (9.78) | 96.62 (0.93) | |
| 16.09 (2.12) | 13.26 (0.75) | 1.64 (0.35) | 104 | 79 | 75 | 3.70 (0.40) | 3.17 (0.67) | 2.50 | 116.79 (2.32) | 92.43 (0.83) | 77.81 (1.39) | 175.56 (12.50) |
| 85.73 (12.79) | 47.16 (13.70) | 1.72 (0.28) | 91 | 87 | 81 | 4.30 | 2.00 | 2.00 | 179.03 (12.79) | 134.16 (13.70) | 87.72 (1.48) | |

*(continued)*

**Table 5.3** (continued)

| Stand of the trees | | Aboveground biomass (t ha$^{-1}$) | | | Underground biomass (t ha$^{-1}$) | | | Total above and underground biomass (t ha$^{-1}$) | | |
|---|---|---|---|---|---|---|---|---|---|---|
| Plantation species | Age | Bottom | Middle | Top | Bottom | Middle | Top | Bottom | Middle | Top |
| 13[f3] | 231.75 (68.28) | 240.85 (73.58) | 211.66 (49.30) | 34.76 (10.24) | 36.13 (11.04) | 31.75 (7.40) | 266.51 (78.52) | 276.98 (84.61) | 243.41 (56.70) | |
| 23[f4] | 202.55 (39.48) | 220.80 (28.54) | 140.38 (17.26) | 30.38 (5.92) | 33.12 (4.28) | 21.06 (2.59) | 232.93 (45.40) | 253.92 (32.83) | 161.44 (19.85) | |
| *E. camaldulensis*[g] | 08[g1] | 289.99 (152.40) | 88.84 (18.47) | 72.21 (3.83) | 43.50 (22.86) | 13.33 (2.77) | 10.83 (0.58) | 333.49 (175.25) | 102.16 (21.24) | 83.04 (4.41) |
| | 18[g2] | 500.72 (183.34) | 138.05 (17.94) | 44.79 (18.73) | 75.11 (27.50) | 20.71 (2.69) | 6.72 (2.81) | 210.85 (210.85) | 158.76 (20.63) | 51.51 (21.54) |
| *G. arborea*[h] | 13 | 143.17 (1.81) | 113.93 (0.82) | 51.79 (26.91) | 21.48 (0.27) | 17.09 (0.12) | 7.77 (4.04) | 164.65 (2.08) | 131.02 (0.94) | 59.56 (30.95) |
| *L. speciosa*[i] | 18[i1] | 570.04 (18.73) | 125.77 (23.24) | 104.70 (32.10) | 85.51 (2.81) | 18.87 (3.49) | 15.71 (4.81) | 655.55 (21.54) | 144.64 (26.73) | 120.41 (36.91) |
| | 20[i2] | 480.89 (32.25) | 28.15 (3.90) | 58.45 (9.01) | 72.13 (4.84) | 4.22 (0.58) | 8.77 (1.35) | 553.02 (37.09) | 32.37 (4.48) | 67.21 (10.36) |
| *P. caribaea*[j] | 18 | 267.59 (30.07) | 126.22 (6.27) | 134.20 (3.61) | 40.14 (4.51) | 18.93 (0.94) | 20.13 (0.54) | 307.73 (34.58) | 145.15 (7.22) | 154.33 (4.15) |
| *S. mahagoni*[k] | 11[k1] | 104.51 (36.29) | 37.76 (5.88) | 30.60 (4.68) | 15.68 (5.44) | 5.66 (0.88) | 4.59 (0.70) | 120.19 (41.73) | 43.42 (6.76) | 35.19 (5.39) |
| | 12[k2] | 153.55 (63.74) | 17.73 (2.07) | 10.60 (1.33) | 23.03 (9.56) | 2.66 (0.31) | 1.59 (0.20) | 176.58 (73.30) | 20.39 (2.38) | 12.19 (1.53) |
| | 15[k3] | 156.69 (10.65) | 128.76 (9.27) | 39.42 (31.56) | 23.50 (1.60) | 19.31 (1.39) | 5.91 (4.73) | 180.20 (12.24) | 148.07 (10.67) | 45.33 (36.29) |
| *S. grande*[l] | 13 | 119.33 (8.21) | 110.78 (22.58) | 52.35 (6.87) | 17.90 (1.23) | 16.62 (3.39) | 7.85 (1.03) | 137.23 (9.45) | 127.40 (25.97) | 60.20 (7.90) |
| *T. grandism* | 20 | 315.96 (54.74) | 208.83 (66.77) | 105.20 (23.52) | 47.39 (8.21) | 31.32 (10.02) | 15.78 (3.53) | 363.35 (62.95) | 240.15 (76.79) | 120.98 (27.04) |
| Slope average | | 264 (19.18) | 130 (8.29) | 72.39 (8.9) | 39.6 (2.9) | 19.51 (1.24) | 10.86 (1.34) | 303.6 (22.1) | 149.61 (9.54) | 83.25 (10.24) |
| Variable average | | 155.50 (9.08) | | | 23.32 (1.36) | | | 178.82 (10.45) | | |

[1]a1a3a6a7; [2]abcefghijk1m; [3]afhk1
Figure in parenthesis indicates the SE
The key of the smaller category appears under the category with larger mean at $p < 0.05$ level

| Total carbon contents of above and underground biomass (tC $ha^{-1}$) | | | Soil carbon contents (tC $ha^{-1}$) | | | Fallen litter carbon contents (tC $ha^{-1} yr^{-1}$) | | | Total carbon contents (tC $ha^{-1}$) | | | |
|---|---|---|---|---|---|---|---|---|---|---|---|---|
| Bottom | Middle | Top | Bottom | Middle | Top | Bottom | Middle | Top | Bottom mean | Middle | Top | Species |
| 133.25 (39.26) | 138.49 (42.31) | 121.70 (28.35) | 113.93 | 94 | 90 | 2.70 | 2.70 | 2.70 | 247.95 (39.26) | 233.19[f2f3] (42.31) | 212.40 (28.35) | |
| 116.46 (22.70) | 126.96 (16.41) | 80.72 (9.93) | 143 | 136 | 120 | 1.70 | 1.70 | 1.70 | 260.16 (22.70) | 263.66 (16.41) | 201.42 (9.93) | |
| 166.75 (87.63) | 51.08 (10.62) | 41.52 (2.20) | 91 | 79 | 75 | 2.80 | 2.80 | 2.80 | 260.55 (87.63) | 132.88 (10.62) | 119.32 (2.20) | 202.56 (31.45) |
| 287.92 (105.42) | 79.38 (10.31) | 25.76 (10.77) | 112 | 98 | 91 | 2.93 (0.13) | 2.80 | 2.80 | 402.85 (105.33) | 180.18 (10.31) | 119.56 (10.77) | |
| 82.32 (1.04) | 65.51 (0.47) | 29.78 (15.47) | 106 | 89 | 72 | 3 | 3 | 3 | 191.32 (1.04) | 157.51 (0.47) | 104.78 (15.47) | 151.20 (13.36) |
| 327.77 (10.77) | 72.32 (13.36) | 60.20 (18.46) | 116 | 107 | 94 | 2.60 | 2.60 | 2.60 | 446.37 (10.77) | 181.92 (13.36) | 156.80 (18.46) | 247.10[3] |
| 276.51 (18.55) | 16.19 (2.24) | 33.61 (5.18) | 134 | 117 | 110 | 3.40 | 3.40 | 3.40 | 413.91 (18.55) | 136.59 (2.24) | 147.01 (5.18) | (31.96) |
| 153.86 (17.29) | 72.58 (3.61) | 77.17 (2.07) | 95 | 83 | 71 | 2 | 2 | 2 | 250.86 (17.29) | 157.58 (3.61) | 150.17 (2.07) | 186.20 (16.99) |
| 60.10 (20.87) | 21.71 (3.38) | 17.59 (2.69) | 100 | 94 | 78 | 1.80 | 1.80 | 1.80 | 160.35 (19.55) | 114.51 (3.38) | 97.39 (2.69) | 140.47 (8.99) |
| 88.29 (36.65) | 10.19 (1.19) | 6.10 (0.76) | 102 | 92 | 88 | 2.10 | 2.10 | 2.10 | 192.39 (36.65) | 104.29 (1.19) | 96.20 (0.76) | |
| 90.10 (6.12) | 74.04 (5.33) | 22.67 (18.15) | 113 | 98 | 95 | 2.10 | 2.10 | 2.10 | 205.20 (6.12) | 174.14 (5.33) | 119.77 (18.15) | |
| 68.62 (4.72) | 63.70 (12.98) | 30.10 (3.95) | 104 | 84 | 64 | 2 | 2 | 2 | 175.62 (4.72) | 149.70 (12.98) | 96.10 (3.95) | 140.47 (12.42) |
| 181.68 (31.48) | 120.08 (38.40) | 60.49 (13.52) | 110 | 100 | 91 | 3.00 (0.36) | 3.00 (0.36) | 3.00 (0.36) | 297.21 (31.67) | 223.08 (38.20) | 151.49 (13.92) | 223.92 (25.76) |
| 151.79 (11.03) | 74.81 (4.77) | 41.62 (5.12) | 108.14 (1.38) | 97.54 (1.39) | 87.37 (1.33) | 2.85 (0.08) | 2.75 (0.08) | 2.76 (0.08) | 262.32 (11.42) | 174.61 (5.44) | 131.61 (5.48) | 189.51 (5.72) |
| 89.41 (5.22) | | | 97.68 (0.95) | | | 2.79 (0.05) | | | 189.51 (5.72) | | | |

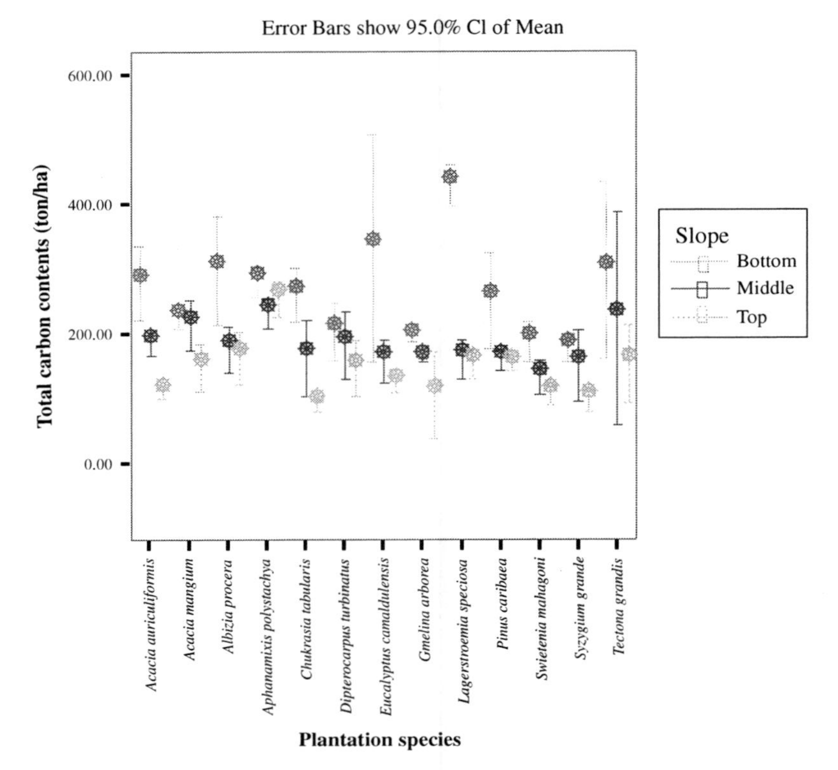


**Fig. 5.1** Total carbon contents by species and slope of the hill in the Chittagong region, Bangladesh

The socio-cultural causes of this higher quantity of fuelwood and litter collection were not explored in this study. Anyway, this study at least shows that these two species in the stands are relatively more sensitive to be reforested under the CDM projects for sequestering carbon. One-way ANOVA between net MAI in carbon stock in species (average) and stands of species showed the significant ($p < 0.05$, $p < 0.01$, and $p < 0.001$) difference.

Miah et al. (2001) studied the 3-year-old stands in the Chittagong hilly areas to determine the net carbon sequestration. In that study, 3 years after establishment of the plantation in the denuded hills of Chittagong region, net total carbon was 79.36 tC $ha^{-1}$ in tree tissues and soil. Before plantation, around 30 tC $ha^{-1}$ was observed in the soil and after 3 years the increased net carbon content in the soil was 20 tC $ha^{-1}$ irrespective of hill positions. It can show the baseline for soil carbon contents in the whole Chittagong region, because of the same ecological zone and soil condition. So, the total carbon contents in the soil (Table 5.3) show the clear addition of organic carbon due to reforestation.

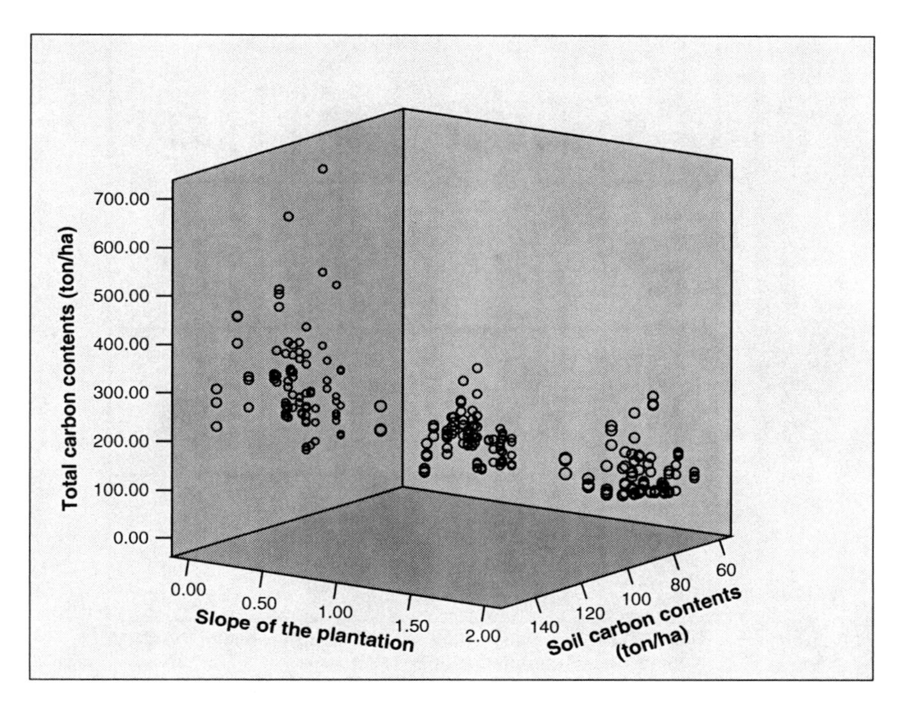


**Fig. 5.2** Relationship between total carbon and soil carbon contents by position of the hill in the Chittagong region, Bangladesh. Notes: 0 = bottom hill, 1 = middle hill, and 2 = top hill

## *5.3.4 Potentialities of Carbon Sequestration in Chittagong Region*

Hill forests in the Chittagong region in Bangladesh lie in Bandarban, Chittagong, Cox's Bazar, Khagrachari, and Rangamati districts. The most important forest in the Chittagong hilly region is the tropical mixed evergreen, which comprises hardwood species, including the highly valued *Dipterocarpus* spp. The total forest areas in the defined Chittagong region is 1,355,288 ha in which 683,430 ha is officially controlled by the Forest Department and the rest is by District Commissioners (pers. commun.[1]). Due to the increasing human dimensions like overextraction and encroachment imposed upon the forests, these have been degraded severely in the meantime. It has been calculated that the closed forests controlled by the Forest Department in that region have dwindled to 40% while the types controlled by the

[1]The data have been collected from the Forestry Directorate, Dhaka, Bangladesh, by the corresponding author himself.

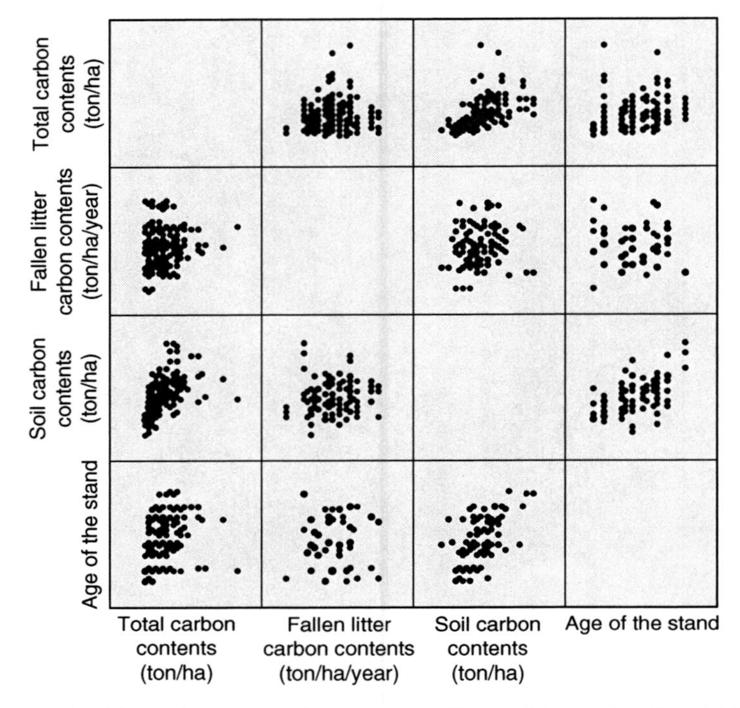


**Fig. 5.3** Relationship matrix of total carbon contents with age of the stand, soil, and fallen litter carbon contents in the Chittagong hilly region, Bangladesh

DC have dwindled to 20%. Hence, a total of 947,545 ha of forestlands have been degraded severely in the whole Chittagong hilly region. In the meantime, around 10% forestlands have been reforested with indigenous and exotic tree species. Therefore, around 94,755 ha of degraded forestlands are still waiting to be replanted in the Chittagong region. Based on the calculations of this study, it can be said that if the whole vacant areas are replanted with the same species then the area may increase the carbon to 379,020 t $yr^{-1}$ (94,755 × 4), which will add the carbon to the stands cumulatively every year, provided there is sustainable forest management.

As the study considered major plantation species with different ages and collection of fuelwood and litters from the stands, it can be assumed that the overall scenarios will be more or less representative for the whole Chittagong hilly region. Most of the species considered in this study has been reported to be selected for plantation in other degraded forest areas in Bangladesh (Hocking & Islam 1998; Kabir & Webb 2005), so the results of the present study for different stands of species with different ages may be useful for the pre-assumption of the carbon sequestration in the other degraded forest zones of Bangladesh.

**Table 5.4** Increment of carbon contents in the stands of different aged stands of tree species in the hilly areas of Chittagong region, Bangladesh

| Stand of the trees | | MAI in tree biomass | MAI in litter biomass | MAI of total carbon in the tree tissues | MALF in carbon | MALL in carbon | MATL of carbon | Net MAI in carbon stock ($tc\ ha^{-1}\ yr^{-1}r$) | |
|---|---|---|---|---|---|---|---|---|---|
| Species | Age (yrs) | ($t\ ha^{-1}\ yr^{-1}$) | ($t\ ha^{-1}\ yr^{-1}$) | ($tc\ ha^{-1}yr^{-1}$) | ($tc\ ha^{-1}yr^{-1}$) | ($tc\ ha^{-1}\ yr^{-1}$) | ($tc\ ha^{-1}\ yr^{-1}$) | Stand | Species |
| *A. auriculiformis* | 06 | 7.43 (1.33) | 5.53 (0.56) | 6.48 (0.75) | 2.84 (0.35) | 0.84 | 3.68 (0.07) | 2.73 (0.70) | 5.81 (0.72) |
| | 08 | 19.73 (6.05) | 6.67 (0.28) | 13.20 (3.06) | 2.84 (0.35) | 0.84 | 3.68 (0.07) | 9.52 (3.06) | |
| | 11 | 11.88 (2.19) | 5.60 (0.24) | 10.21 (2.13) | 2.84 (0.35) | 0.84 | 3.68 (0.07) | 6.53 (2.13) | |
| | 15 | 13.58 (4.17) | 5.25 (0.22) | 9.42 (2.13) | 2.84 (0.35) | 0.84 | 3.68 (0.07) | 5.74 (2.13) | |
| | 16 | 13.68 (3.51) | 6.36 (0.57) | 10.02 (1.89) | 2.84 (0.35) | 0.84 | 3.68 (0.07) | 6.34 (1.89) | |
| | 17 | 9.96 (2.52) | 7.91 (0.40) | 8.94 (1.18) | 2.84 (0.35) | 0.84 | 3.68 (0.07) | 5.26 (1.18) | |
| | 18 | 7.91 (1.56) | 8.60 | 8.25 (0.78) | 2.84 (0.35) | 0.84 | 3.68 (0.07) | 4.57 (0.78) | |
| *A. mangium* | 11 | 14.94 (1.31) | 5.22 (0.15) | 10.08 (0.68) | 1.70 (0.1) | 0.82 | 2.52 | 7.48 (0.66) | 7.48 (0.66) |
| *A. procera* | 20 | 9.01 (1.55) | 5.11 (0.11) | 7.06 (0.77) | 2.28 (0.21) | 0.41 | 2.69 | 4.31 (0.74) | 4.31 (0.74) |
| *A. polystachya* | 12 | 25.39 (1.03) | 5.89 (0.31) | 15.64 (0.53) | 9.02 (1.2) | 1.84 | 10.86 | 4.38 (0.39) | 2.15 (0.60) |
| | 13 | 14.54 (0.70) | 7.00 | 10.77 (0.35) | 9.02 (1.2) | 1.84 | 10.86 | −0.09 (0.35) | |
| *C. tabularis* | 08 | 10.64 (4.24) | 5.70 (0.11) | 8.17 (2.15) | 2.58 (0.13) | 0.60 | 3.18 | 4.73 (2.01) | 3.82 (1.09) |
| | 17 | 7.68 (1.85) | 4.51 (0.11) | 6.09 (0.90) | 2.58 (0.13) | 0.60 | 3.18 | 2.91 (0.90) | |
| *D. turbinatus* | 06 | 3.44 (0.77) | 8.58 (0.68) | 6.01 (0.63) | 3.45 (0.23) | 0.94 | 4.39 (0.04) | 1.94 (0.68) | 2.62 (0.43) |
| | 08 | 7.33 (2.06) | 5.53 (0.77) | 6.43 (1.33) | 3.45 (0.23) | 0.94 | 4.39 (0.04) | 2.04 (1.33) | |
| | 13 | 12.40 (0.89) | 5.40 | 8.90 (0.45) | 3.45 (0.23) | 0.94 | 4.39 (0.04) | 4.51 (0.45) | |
| | 23 | 9.40 (0.96) | 3.40 | 6.40 (0.48) | 3.45 (0.23) | 0.94 | 4.39 (0.04) | 2.01 (0.48) | |
| *E. canaldulensis* | 08 | 15.61 (2.58) | 5.60 | 10.89 (1.30) | 0.46 | 0.21 | 0.67 | 10.22 (1.30) | 9.83 (1.50) |
| | 18 | 14.56 (5.60) | 5.69 (0.09) | 10.12 (2.80) | 0.46 | 0.21 | 0.67 | 9.45 (2.80) | |
| *G. arborea* | 13 | 8.33 (1.41) | 5.00 | 6.559 (0.67) | 3.65 (0.5) | 2.60 | 6.25 | 0.25 (0.64) | 0.25 (0.64) |
| *L. speciosa* | 18 | 14.83 (3.96) | 6.31 (0.20) | 10.90 (1.84) | 6.23 (0.29) | 2.58 | 8.81 | 2.09 (1.84) | 0.70 (1.30) |
| | 20 | 9.21 (3.63) | 6.84 (0.04) | 8.03 (1.82) | 6.23 (0.29) | 2.58 | 8.81 | −0.68 (1.84) | |
| *P. caribaea* | 18 | 10.13 (1.03) | 4.00 | 7.07 (0.51) | 0.86 | 0.48 | 1.34 | 5.66 (0.46) | 5.66 (0.46) |
| *S. mahagoni* | 11 | 6.02 (1.66) | 3.60 | 4.81 (0.83) | 2.39 (0.4) | 0.83 | 3.22 (0.03) | 1.56 (0.83) | 1.77 (0.47) |
| | 12 | 3.81 (1.54) | 4.20 | 4.01 (0.77) | 2.39 (0.4) | 0.83 | 3.22 (0.03) | 0.73 (0.73) | |
| | 15 | 8.30 (1.56) | 4.20 | 6.25 (0.78) | 2.39 (0.4) | 0.83 | 3.22 (0.03) | 3.03 (0.78) | |
| *S. grande* | 13 | 7.44 (0.81) | 4.00 | 5.50 (0.34) | 3.20 (0.43) | 0.56 | 3.76 (0.23) | 1.74 (0.34) | 1.74 (0.34) |
| *T. grandis* | 20 | 12.07 (2.29) | 6.87 (0.43) | 9.47 (1.16) | 2.03 (0.21) | 3.42 | 5.45 (0.12) | 4.05 (1.15) | 4.05 (1.15) |
| *Average* | | 11.05 (0.56) | 5.66 (0.10) | 8.42 (0.30) | 3.26 (0.13) | 1.10 (0.05) | 4.36 (0.16) | 4.04 (0.31) | |

MAI = mean annual increment; MALL = mean annual loss through litter
MALF = mean annual loss through fuelwood; MATL = mean annual total loss
Figure in parenthesis indicates the SE

## 5.4 Global Warming and Its Effects on Bangladesh

At present, global warming is a matter of grave concern. Since the late 19th century, global temperature has increased by 0.3–0.6°C, and, globally, sea levels have risen 10–15 cm over the past 100 years (IPCC 2001a). Due to overpopulation, especially in Asian and African countries, natural resources are under extreme pressure, which, cumulatively, causes environmental problems. Mayaux et al. (2005) and Achard et al. (2002) reported that world humid tropical forests had been disappearing at a rate of about 5.8 (±1.4) m ha $yr^{-1}$, with a further 2.3 (±0.7) m ha $yr^{-1}$ of forests visibly degraded between 1990 and 1997. Soil degradation affects more than 900 m people in 100 countries, and about 1.1 billion rural dwellers are at risk from desertification and dry-land degradation. Natural and man-made disasters, in the last couple of decades, have maintained a high frequency. If these short-term climatic variations are actually related to long-term climate change, they may provide terrifying examples of what the future holds if the world fails to curtail its $CO_2$ emissions. Changes in climate will affect coastal systems through sea-level rise (projected to increase by 15–90 cm by 2100) and an increase in storm surge hazards and possible changes in the frequency and/or intensities of extreme events. Bangladesh could be a major victim of all these unfavorable consequences of GHG accumulation, as it has limited scope to develop preventive measures.

## 5.5 CDM Additionality in Bangladesh

Bangladesh is a poor developing country in South Asia. Massive reforestation programs in Bangladesh started from the 1980s. But almost all of the plantation programs were implemented with the aid of donor agencies (pers. commun.). Without those aids, plantation could not be established. Major causes of deforestation of natural forests or plantation are increasing population and poverty in Bangladesh (FAO 1998). Figure 4.5 shows the annual growth trend of population and GDP from 1985 to 2003 (SESRTCIC 2006). Considering the status quo of the present population, GDP, and poverty rate described in Sect. 4.6, it can be stated that still Bangladesh cannot establish the plantation with her own fund and technology. Furthermore, there is no more expectation to have aid for reforestation in Bangladesh. So, without the CDM project, reforestation is not possible in the country, which shows the existence of 'additionality' of CDM in Bangladesh.

## 5.6 Prices for Carbon Credits

The cost of carbon sequestration varies from region to region and also from country to country, based on different economic analyses. Kim et al. (2004) estimated the cost of carbon at around US $19.7 per Mg C in Southeast Asian countries. Kirschbaum (2001) assumed a cost of US $10 per Mg C for indefinite carbon savings in different arbitrary accounting periods. Missfeldt & Haites (2001) used a 1995 cost of US $15 per Mg C for a sink enhancement scenario, and Tschakert (2004) used a cost of US $15 per Mg C for her study in Senegal. In other studies, it ranged from US $1 to 100 (Healey et al. 2000; Missfeldt & Haites 2001; Niles et al. 2002; Roper 2001). There is no study on prices of carbon credits in Bangladesh but it can be assumed from the above findings that the price would range from US $15 to 20 per Mg C based on the scenarios on Southeast Asian countries and Senegal assuming the same socio-economic condition.

### *5.6.1 Carbon Accounting Method*

Carbon accounting method, 'tonne-year,' may be appropriate for carbon trading through the CDM forestry project in Bangladesh, which was outlined by Marland et al. (2001) and Tipper & de Jong (1998). Tonne-years combine the quantity of carbon sequestered in a project with the longevity of the project. It was originally proposed as an effective way to prevent deforestation: it rewards those who preserve forests in proportion to both the amount of the carbon held and the duration for which they held it. The existing wood volumes and soil organic matters per unit area for individual years during the project period should be predicted by collecting growth data on the diameter at breast height (dbh) together with the height of a tree and the density of alive trees. This prediction can be done by applying these growth data to the approximation curve (for instance, the Mitscherlich equation, logistic equation, and nonlinear regression equation). It is also necessary that carbon credits be allocated based on the net annual increment in carbon in each stand. However, credits are allocated based on the net quantity of carbon sequestration, in which the felled quantity is deducted from the growth of all stands.

## 5.7 Sustainability of the CDM Project in Bangladesh

People of rural areas depend mainly on traditional fuels, namely wood, and agricultural residues for domestic consumption. Fuelwood makes up around 13% of the whole biomass energy consumption of the country and 84% of total roundwood production (GOB 1993). Fuelwood is the dominant domestic fuel in both rural and urban areas. About one-half of the wood fuel is used for cooking, about one-third for heating the house, boiling water, etc., and the remainder for other domestic purposes, such as agricultural processing and industry. Fuelwood collection from the forests is thought to be one of the major causes of deforestation in Bangladesh. It may hamper the investor's decision regarding forestry-offset projects in Bangladesh. So, to maintain the sustainability of the CDM projects, it is important that they are designed in harmony with the supply and demand for fuelwood. A sustainable forest management approach, with the possibility of fuelwood substitution, may resolve this problem. Therefore, the balance between the fuelwood production and its uses by people and the availability of substitutes may lead to the success of the CDM projects. Research on elasticity of the consumption of fuelwood may assist the decisions of policy makers regarding fuelwood substitution. However, to reduce the transaction costs of CDM projects, the design of proper incentive contracts could result in a possible solution for the investor, host country, and the environment.

Sustainability of the user groups in the community plantation programs is critical to the success of the CDM program. However, the size of the user group, the sharing system, heterogeneity of endowments, homogeneity of identities and interests, fairness in allocation of benefits from forest resources, level of user demand and institutional characteristics, like the relative simplicity and ease in understanding the rules, the access, and management rules and sanctions determine the sustainability of the user groups under the CDM (Gundimeda 2004). Agrawal (2001) also made similar comments for the sustainability of user groups in the community plantation programs.

The CDM of the Kyoto Protocol is a dual-purpose mechanism to enable industrialized countries both to reduce emission reduction costs and also to contribute to sustainable development in developing countries (Smith 2002b). The COP7 Agreements opened the door to proactive measures in promoting sustainable development by giving the SBSTA the responsibility for developing modalities to address the environmental and social implications of reforestation and afforestation projects. However, they also limit the extent to which proactive measures can be taken at an international level, by allowing national governments to decide whether or not a project contributes to sustainable development (Smith 2002b).

Commercial plantations are likely to dominate and contribute toward lowering the cost of emission reduction in Bangladesh. However, there may be associated social risks that need to be addressed, such as the possibility of commercial plantations dispossessing local communities of their land and livelihoods. These risks are likely to be most serious where land conflicts are endemic and rural land tenure is unclear and overlapping, a situation that prevails in Bangladesh. In this regard, Smith (2002b) emphasized the political dimension, i.e., that social risks are magnified where government regimes are repressive, governance is poor, and strong economic and political alliances exist between governments and the timber industry. The political dimension in Bangladesh is simply characterized by Smith's analysis (Smith 2002b). These social risks are unlikely to be fully addressed, given the host country's sovereignty over the determination of sustainable development. Social and environmental risks are likely to be significantly lower for community-based plantations. The agreements on simplified baselines and monitoring for small-scale projects with community plantations will be particularly beneficial to smallholders.

## 5.8 Problems and Constraints for Creating CDM Forests in Bangladesh

The most important constraint to Bangladesh's participation in carbon trading is the lack of capacities to deal with the CDM projects. The major issues include baselines and additionality, leakage, and monitoring. Other important issues are sustainable development as the major objective and the sustainability of the CDM project. In addition, there is insufficient research on understanding the carbon sink in the various forest areas of Bangladesh.

The inadequacy of net emission estimations from forests is one of the criticisms of forestry-based compliance options, which is creating a contentious debate in the land-use mitigation options (LeBlanc 1999). The estimates for national or regional land-use change emission inventories are often large because of uncertainties about deforestation rates, the fate of land after deforestation, and the amount of biomass in different ecosystems over large land areas (IPCC 1996). Some studies estimate annual carbon emissions, from land-use change in the tropics, to be 1.2–2.2 (IPCC 2000) or 1.5–3.0 b t (IPCC 1992). However, carbon storage, at the project level or on discrete parcels of managed land, can be estimated with a much smaller standard error because many of the underlying uncertainties, such as condition of the land and type of land use, are resolved (LeBlanc 1999). In addition to this, carbon estimation can be easily performed in the forests, which are managed scientifically, where biomass can be estimated by some allometric equations by simple measurements of height and diameter. Developing countries, like Bangladesh, may have some reservations on forestry-offset projects which may conflict with national sovereignty due to the presence of foreign investors, which basically arise from the long gestation period of the CDM projects, leading to the long-term dictation of land-use patterns (LeBlanc 1999).

Most of the plantations in Bangladesh have been established by community participation in the encroached forest areas in Bangladesh. In future it also may be expected that plantation will be established by community participation. But it may have uncertainty of carbon credit due to some uncertainties in the community participation in establishing plantations. Interest in tree planting develops under certain rather limited conditions. Insecure land tenure, unfavorable policies such as bureaucratic restrictions, and inter-sectoral conflicts may hamper the community-level plantation establishment (Khan 1998). Higher opportunity cost of land on small farms may increase the cost of carbon sequestration by community plantation (Smith 2002b). Higher transaction costs may also discourage community plantation (Smith 2002b). If the population growth cannot be reduced and vertical development in the agricultural sector does not occur, then CDM forests will obviously conflict with the agricultural sector. The ever-increasing trend of urbanization and industrialization also will conflict with the CDM forests. In this case only proper policy can attract the investors for the CDM projects in Bangladesh.

## 5.9 Conclusions

This chapter finds that reforestation makes a significant contribution to carbon sequestration generating CER in Bangladesh. It is also expected that much revenues can be earned by selling CER in the carbon market through CDM projects. The comparatively lower cost of the carbon sequestration in Bangladesh can attract the investors of the Annex I countries. Forestry lands, used under the CDM, would accrue the benefits to the national forestry sector, as well as the private owners and participants in the community forestry, in terms of an overall increase in income and to achieving self-sufficiency. On the other hand, there may be conflicts, as the short-term needs of the poor, especially for fuelwood and timber, differ with the long-term requirement of carbon storage. Inter-sectoral conflicts also may emerge during the implementation of the CDM projects. Appropriate economic institutions and mechanisms need to be established for the CDM to result in equity and sustainable development within the country. CDM projects should be designed in such a way that they evolve benefit, not only for local people but also for the sustainability of the project itself.

The gross carbon content in the forests of Bangladesh, in general, and in the Chittagong hilly area, in particular, indicates that Bangladesh has a high capacity of carbon sequestration. Due to adverse human dimensions like overextraction of the forest resources, on the plantations, however, the net annual carbon increment is lessened. Nevertheless, the content range of organic carbon in the stand soils clearly shows the significant potential of the Chittagong hilly lands to sequester carbon. Literature reviews on gross carbon content in other parts of Bangladesh, and in the different forests and soils of India, show that the stands in Chittagong have a comparatively higher carbon sequestration capacity. Biomass collection from the forests of the Chittagong area shows the high local preference for particular tree species in terms of fuel use. The net MAI of carbon stock in the plantations of different species also indicates the potentiality of carbon sequestration in the expected plantations in the Chittagong region. The results of the study emphasize that every reforestation program in this area should address the needs of the local people, thereby leading to sustainability of the forests, while also sustaining the carbon increment and sequestration. The results may be an important tool in managing plantations more efficiently with minimum carbon loss. The study clearly shows the capacity of some plantations to sequester atmospheric carbon. This may, in turn, be representative of the capability of the hundreds of thousands of hectares of expected CDM forests in the Chittagong region as well as in the whole country to sequester huge quantities of carbon from the atmosphere, through which Bangladesh may claim payment for reducing global warming.

The effects of global warming on Bangladesh, as discussed in this study, show serious consequences for the economy. Even though the 'effects' are not created by Bangladesh itself, the country could suffer enormously from any possible climate change, due to its severe poverty and lack of preparedness to combat the hazards. The participation of Bangladesh as a Non-Annex I country in the global warming mitigation program involves sharing with the international efforts to reduce global warming and developing the forestry sector, as well as to build up the capacity to combat the likely natural hazard due to uncontrolled climate change. To gain the sustainability of the CDM projects in the country, the study recommends harmonizing the primary fuelwood energy needs to the industrial biomass conversion options and emphasizing on the sustainability of the user groups. The study finds the most important constraint to Bangladesh's participation in carbon trading as the lack of capacities to deal with the CDM projects. So, capacity building should be prioritized to host the CDM projects in the country as early as possible.

The quantification of carbon sequestration, by the present study, can direct policy makers, researchers, and administrators in bargaining the price of international GHG reduction, which can advance the economic, social, and environmental development of Bangladesh.

# Chapter 6
# Conclusions and Recommendations

## 6.1 Conclusions

Cost-effective reduction of GHGs is an important task in the arena of climate change mitigation. This book importantly discusses and argues forests to climate change mitigation as one of many positive response options. It finds both in abstraction and Bangladesh in particular, that forests/forestlands are potential to reduce global warming which can mitigate global climate change effectively. The CDM, an important economic mechanism formulated by the Kyoto Protocol, has been focused as an effective tool for GHG reduction and sustainable development in the non-Annex I countries. As a case study in Bangladesh, this book shows a higher capacity of forests to sequester atmospheric carbon. Although, CDM only allows A/R activities in the first commitment period of the Kyoto Protocol, this book argues that reducing emissions from deforestation and degradation is also promising for climate change mitigation and positive for short-term 'sustainable development effect' in the non-Annex I countries. Fuel substitution and energy efficiency-based CDM projects can gain momentum in the country. Socio-economic impacts of biomass energy promotion in Bangladesh show that it can provide social cohesion and stability and increased standard of living of the people, spreading positive flow to the whole economy of the country. Thus, the country can achieve sustainable development. In the production arena of the forest biomass, the study proves that Bangladesh has enormous potential to produce biomass in blank, encroached, or other degraded forestlands and homestead compounds. To gain sustainability of the CDM projects in the country, the study recommends harmonizing the primary fuelwood energy needs to the industrial biomass conversion options and emphasizing on the sustainability of the user groups. The study finds that the most important constraint to Bangladesh's participation in carbon trading is the lack of capacity to deal with the CDM projects. So, capacity building should be prioritized to host the CDM projects in the country as early as possible.

The discussion on the reforestation success in the Republic of Korea made clear that rapid poverty alleviation, spontaneous mass participation, and political commitment acted as a mainstream to reforest the degraded forestlands effectively.

Md.D. Miah et al., *Forests to Climate Change Mitigation*, Environmental Science and Engineering, DOI 10.1007/978-3-642-13253-7_6, © Springer-Verlag Berlin Heidelberg 2011

The study indicates that improvement of the traditional cooking stove can save around 50% fuel and cooking time as well as GHG emissions in Bangladesh. The improved cooking stoves also can improve the kitchen environment and save women and children from corresponding health hazards.

## 6.2 Recommendations

### *6.2.1 To Reorient the General Future Policy/Approaches of the Forestry Sector*

The future policy/approach of the forestry sector should be oriented to retard the deforestation, reforest the degraded forestlands, and afforest the newly accredited lands. The general objectives of the forestry sector should be the conservation of biodiversity, the mitigation of global warming, and the alleviation of poverty. To reorient the forestry sector in that way, both physical and institutional measures are needed. Development of coastal green belts, agroforestry, and social forestry may be included in the physical measures. Institutional measures may include integrated ecosystem planning and management, management of ecosystem in the reserved/protected areas, and reduction of habitat fragmentation. The participatory forestry approach should be reoriented with proper benefit sharing by the local participants with clear land tenure and keeping the idea of sustainability and biodiversity.

Making political commitment is critical to both preserving the forests and eliminating the corruption in the forestry sector of Bangladesh. FAO (2001) addressed the illegal practices in the forest, i.e., illegal occupation of forestlands, illegal logging, illegal timber transport, trade and timber smuggling, transfer pricing and other illegal accounting practices, and illegal forest processing. To reduce these illegal activities, the following measures may be undertaken: (i) providing rewards to the foresters for integrity; (ii) increasing the probability of detection of forest crime and detection; (iii) increasing the penalties for forest crime and corruption; (iv) reducing the discretionary power of government officials; (v) increasing the use of market mechanisms; (vi) involving the media, NGOs, and the public in combating forest crime; and (vii) reinforcing the forest laws and legislation very strictly.

A comprehensive awareness program should be extended to the general people, private forest owners, local and national politicians, and so forth, aimed at showing the importance of forestry, environmental conservation, biodiversity, and endangered ecosystems. Adequate and proper training should be provided to the officials in charge of enforcing various policies and regulations.

The government has recently promulgated the National Forest Policy (1994) and the National Energy Policy (1995). Both of these policies, for the first time, specifically address the issue of woodfuel and have pledged government commitment to ameliorate the situation. The Forest Policy aims to increase the tree cover area up to 20% by the year 2015 through both government and private sector efforts. The Forest Policy further mentions that forests will be extended to rural areas, newly accreted chars, USF lands, and other fallow private and community lands such as roadsides, railway sides, institution premises, embankment slopes, and homesteads by public participation. The Energy Policy addresses the energy issues from the viewpoints of current status and future programs under renewable and non-renewable energy sources.

It emphasizes the conservation of biomass fuel by introducing fuel-saving technologies like improved cooking stoves and their dissemination to both rural and urban households and commercial installations. Hence motivation has to be strengthened. The statements on biomass conversion to other energies are not clear. Although the Government of Bangladesh is motivated to promote biomass energy with biased statements, it has not been fully approached to meet the criteria of CDM. So, the energy policy and national forest policy should be reviewed again to comply with the CDM framework.

To reorient the forestry sector of Bangladesh to achieve the 'CER' through CDM, the following general recommendations have been made.

### 6.2.2 *To Achieve the Certified Emission Reduction (CER) Under CDM*

Benítez et al. (2007) asserted that while investing into the Non-Annex I countries in the CDM projects, country particularities like institutions, government credibility, corruption, economic stability, inflation, and terrorism must be considered as political, financial, and economic risk factors. Bangladesh is overwhelmed by these risk factors (Zafarullah & Siddiquee 2001), which may lead to the negative investment trend in the CDM projects. So, Bangladesh should take appropriate policy measures and actions to tackle these risk factors.

The inter-sectoral conflicts among forestry, agriculture, environment, land, wildlife, and energy sectors should be resolved. There is a serious gap in terms of the coordination among economic and environmental objectives in Bangladesh. The gap is more serious in the case of the understanding and coordination of the linkages between GHG abatement activities and measures. Although Bangladesh emits less than 0.1% of global GHG emissions, she is one of the countries that would suffer adverse impacts from anthropogenic climate change (Huq 2001). To cope with the changes of climate, the country needs to develop a concerted plan of action to face the problems of climate change and the development challenges they will present. Huq (2001) asserted that this would require a well-coordinated policy for scientific research and development, focusing particularly on building an adaptive capacity. In particular, such a capacity needs to be developed in the fields of disaster management, agriculture, water resource management, and coastal zone management. The elements of the strategy specific to climate change also need to be incorporated into the national and sectoral planning to ensure that they are compatible with national sustainable development objectives. The policies should minimize the land-use and resource-management regimes. Issues relating to project permanence, leakage, and transaction costs should also be addressed (Kennett 2002).

Bangladesh had repeatedly increased the trend of foreign investment in the industrial sector. But no investment was seen in the forestry sector. A part of the national degraded forestlands can be allotted to the national or multinational companies as a long-term lease for reforestation. Appropriate policies, benefit-sharing mechanisms between the government and the companies, and share of the local communities should be confirmed first in this regard. It could be a better alternative to gain the capability to deal with the CDM forestry activities.

### 6.2.3 To Promote Biomass Production and Its Energy

To promote biomass and its energy in Bangladesh under the framework of CDM, some measures have been recommended hereunder:

*Technical measures.* As land is a scarce resource in Bangladesh, biomass productivity enhancement is mandatory. Likewise, the long rotation period needs to be reduced. Technology information package for high yields should be made accessible to farmers and village communities. Species, type of biomass, and technology-specific GHG mitigation models should be developed.

*Economic measures.* Market or demand creation for biomass energy is critical to the biomass promotion in the country. The financial analysis of Bhattacharya et al. (2003) shows that the benefit–cost ratio of biomass production is positive and thus profitable even at low-to-moderate productivity. Standard rules and procedures, for defining and quantifying carbon credits at the project level, must be developed to achieve the carbon credit under CDM. Balancing the short-term needs of the poor with the long-term requirement for carbon is necessary. It should be ensured that maximum benefit generated from the CDM project should be channeled to the poor. These measures will contribute to enhanced biomass production and its conversion.

*Institutional measures.* Secure land tenures to enable long-term contractual supply of biomass feedstock to bioenergy utilities are necessary. Simplified procedures are needed to promote agreements between farmers and utilities for sustained supply of biomass feedstock. Creating marketing institutions to link biomass producers and bioenergy utilities is important. Creating forestry extension service to provide information to biomass producers will secure the biomass productivity. A strong long-term political commitment to biomass energy promotion is critical to the biomass-based CDM projects in Bangladesh. Minimum standards for stakeholder consultations would go a long way toward addressing social risks and also be politically feasible, given that stakeholder consultations are mandatory for all CDM projects. In the long run, biomass production can be best promoted through promotion of private forestry, which should be treated as similar to agricultural crop production. Therefore, conditions should be created to encourage involvement of private sector in biomass production. Policies and measures that promote biomass energy technologies will also indirectly promote biomass production by creating demand for biomass fuels.

#### 6.2.3.1 To Promote Improved Cooking Stove/Efficient Biomass Burning

The following recommendations are made in regard to improved cooking stoves and reduction of GHG emissions:

(1) Improved cooking stoves should be introduced throughout the rural areas of Bangladesh with minimum cost so that they are easily accessible to the rural poor.
(2) Plantations of fast-growing fuelwood species like *Acacia* spp., *Albizia* spp. should be expanded in the gaps of village groves.
(3) Alternative renewable energy sources should be accessible to rural women, which may reduce the dependency on wood. These may include solar energy and biogas, for which low-cost efficient cookers should be introduced in rural areas.
(4) National program to subsidize the cost of improved cooking stoves should be expanded. Government or non-government organizations may launch a new marketing system to introduce improved cooking stoves to rural areas.
(5) The dissemination of the improved cooking stove is not enough throughout the country. So, the Government and NGO efforts should be strengthened to disseminate this technology

### *6.2.4 To Be Involved with A/R CDM*

Based on the issues discussed in this study, the following recommendations have been made to resolve the difficulties of Bangladesh's participation in the global warming mitigation program through CDM A/R:

(1) Research should be undertaken to collect data on the quantity, distribution, and partitioning of carbon and any changes taking place over time in the different ecological zones of Bangladesh.
(2) Species and site-specific carbon measurement models should be developed for both indigenous and exotic plant species.
(3) Standard rules and procedures, for defining and quantifying carbon credits at the project level, must be developed to feasibly generate forestry offsets under the CDM.
(4) Minimum standards for stakeholder consultations would go a long way toward addressing social risks and may also be politically feasible, given that stakeholder consultations are mandatory for all CDM projects.
(5) To solve the problems or obstacles for the inclusion of forestry activities in CDM projects, including baselines and additionality, leakage, monitoring, and accounting procedures, project requirements, vis-à-vis all of the above areas, must be standardized, straightforward, and easy to apply to avoid excessive transaction costs.

(6) Gundimeda (2004) recommended three important criteria for the sustainability of CDM projects in developing countries, which may conform to the Bangladeshi perspective. These criteria are the following: (a) balancing the short-term needs of the poor with the long-term requirement for carbon; (b) management of forestlands by the rural poor after the design of the CDM project; and (c) ensuring that maximum revenue is channeled to the poor. The first criterion implies that, while designing the CDM projects, care should be taken to ensure that the annual energy requirement of rural dwellers, met by fuelwood and agricultural residues, is incorporated. The second important issue, in project design, is that concerns of the poor participants in the project should be well addressed.

(7) Regarding the continuance of the CDM in Bangladeshi forests, user groups should be given responsibility for the management of government forests, with proper benefit-sharing mechanisms. Guidelines for codes of conduct and ethics, and other institutional arrangements, should be developed to assist the user groups. Although Bangladesh has some experience of social forestry programs, new location-specific revenue-sharing rules should be designed under the CDM. Further extensive research is required; nevertheless, Bangladesh can achieve sustainable development through appropriate economic policies and practices, which can address all uncertainties regarding the CDM projects.

(8) Bangladesh can make an intergovernmental collaboration with the Government of the Republic of Korea to learn how a reforestation scheme can be successful.

# References

Abedin M. Z. & Quddus M. A. 1990. Household fuel situation, homegardens and agroforestry practices at six agroecologically different locations of Bangladesh. In: *Homestead Plantation and Agroforestry in Bangladesh* (eds M. Z. Abedin, C. K. Lai, and M. O. Ali) pp. 19–54. Bangladesh Agricultural Research Institute (BARI), Joydebpur, Bangladesh.

Achard F., Eva H. D., Stibig H. J., Mayaux P., Gallego J., Richards T., & Malingreau J. P. 2002. Determination of deforestation rates of the world's humid tropical forests. *Science* 297: 999–1002.

ADB (Asian Development Bank). 1998. Asia Least-cost Greenhouse Gas Abatement Strategy (ALGAS): Bangladesh. Asian Development Bank, Manila, The Philippines.

ADB. 2001. Key Indicators of Developing Asian and Pacific Countries, 32nd edn. Oxford University Press, China.

ADB. 2004. Key Indicators 2004: Poverty in Asia: Measurement, Estimates, and Prospects. Asian Development Bank.

Agrawal A. 2001. Common property institutions and sustainable governance of resources. *World Development* 29: 1649–1672.

Aggarwal R. K. & Chandel S. S. 2004. Review of improved cookstoves programme in Western Himalayan state of India. *Biomass and Bioenergy* 27: 131–144.

Ahmed G. U. & Bhuiyan M. K. 1994. Regeneration status in the natural forests of Cox's Bazar Forest Division, Bangladesh. *Annals of Forestry* 2: 103–108.

Ahmed G. U., Newaz M. S., & Temu A. B. 1992. Status of natural regeneration in the denuded hills of Chittagong, Bangladesh. *Commonwealth Forestry Review* 71: 178–185.

Akhter J., Millat-E-Mustafa M., Khan N. A., & Alam M. S. 1999. Household biomass fuel energy situation of a forest rich district of Bangladesh. *Bangladesh Journal of Agriculture* 24: 55–65.

Alam M. S., Haque M. F., Abedin M. Z., & Aktar S. 1990. Homestead trees and household fuel uses in and around the FSR site, Jessore. In: *Homestead Plantation and Agroforestry in Bangladesh* (eds M. Z. Abedin, C. K. Lai, and M. O. Ali) pp. 106–119. Bangladesh Agricultural Research Institute (BARI), Joydebpur, Bangladesh.

Alamgir M., Miah M. D., & Haque S. M. S. 2004. Species composition and functional uses of trees outside forests in Chittagong region, Bangladesh. *Chiang Mai Journal of Science* 31: 63–68.

Alcántara-Ayala I. 2002. Geomorphology, natural hazards, vulnerability and prevention of natural disasters in developing countries. *Geomorphology* 47: 107–124.

Ali M. 2002a. Scientific forestry and forest land use in Bangladesh: A discourse analysis of People's attitudes. *International Forestry Review* 4: 214–222.

Ali M. E. 2002b. Transfer of Sustainable Energy Technology to developing countries as a means of reducing greenhouse gas emission – the case of Bangladesh: Review of relevant literature, Discussion Paper No. 02.08 edn. Department of Applied and International Economics, Massey University, New Zealand.

Ara S., Gafur M. A., & Islam K. R. 1989. Growth and biomass production performances of *Acacia auriculiformis* and *Eucalyptus camaldulensis* reforested in the denuded hilly lands. *The Bangladesh Journal of Botany* 18: 187–195.

Md.D. Miah et al., *Forests to Climate Change Mitigation*, Environmental Science and Engineering, DOI 10.1007/978-3-642-13253-7, © Springer-Verlag Berlin Heidelberg 2011

Arnold J. E. M., Kohlin G., & Persson R. 2006. Woodfuels, livelihoods, and policy interventions: Changing perspectives. *World Development* 34: 596–611.

Baral A. & Guha G. S. 2004. Trees for carbon sequestration or fossil fuel substitution: The issue of cost vs. carbon benefit. *Biomass and Bioenergy* 27: 41–55.

BB (The Bangladesh Bank). 2002. Bangladesh: Some Selected Statistics, Appendices of Annual Report 2000/2001. Dhaka, Bangladesh.

BBS (Bangladesh Bureau of Statistics). 1996. Census of Agriculture 1996. Ministry of Planning, Government of the People's Republic of Bangladesh, Dhaka, Bangladesh.

BBS. 1997. Statistical Yearbook of Bangladesh. Ministry of Planning, Government of the People's Republic of Bangladesh, Dhaka, Bangladesh.

BBS. 2000. Statistical Year Book of Bangladesh. Statistics Division, Ministry of Planning, Government of the People's Republic of Bangladesh, Dhaka, Bangladesh.

Begg K., Parkinson S., Mulugetta Y., Wilkonson R., Doig A. & Anderson T. 2000. Initial evaluation of CDM type projects in developing countries. Final report of DFID project 7305. Centre for Environmental Strategy (CES), University of Surrey.

Benítez P. C., McCallum I., Obersteiner M., & Yamagata Y. 2007. Global potential for carbon sequestration: Geographical distribution, country risk and policy implications. *Ecological Economics* 60: 572–583.

Berglund B. E. 2003. Human impact and climate changes-synchronous events and a causal link? *Quaternary International* 105: 7–12.

Berndes G., Hoogwijk M., & van den Broek R. 2003. The contribution of biomass in the future global energy supply: A review of 17 studies. *Biomass and Bioenergy* 25: 1–28.

Bhattacharya S. C., Albina D. O., & Khaing A. M. 2002. Effects of selected parameters on performance and emission of biomass-fired cookstoves. *Biomass and Bioenergy* 23: 387–395.

Bhattacharya S. C., Attalage R. A., Augustus Leon M., Amur G. Q., Salam P. A., & Thanawat C. 1999. Potential of biomass fuel conservation in selected Asian countries. *Energy Conversion and Management* 40: 1141–1162.

Bhattacharya S. C. & Salam P. A. 2002. Low greenhouse gas biomass options for cooking in the developing countries. *Biomass and Bioenergy* 22: 305–317.

Bhattacharya S. C., Salam P. A., Pham H. L., & Ravindranath N. H. 2003. Sustainable biomass production for energy in selected Asian countries. *Biomass and Bioenergy* 25: 471–482.

Bhuiyan M. K., Haque S. M. S., Hossain M. M., & Ali M. 1995. Exploitation of forest produces from the Chittagong University campus and its adjoining forest areas. *Chittagong University Studies, Part II: Science* 19: 1–5.

Biswas W. K., Bryce P., & Diesendorf M. 2001. Model for empowering rural poor through renewable energy technologies in Bangladesh. *Environmental Science & Policy* 4: 333–344.

BPDB (Bangladesh Power Development Board). 2002. Key Statistics. Bangladesh Power Development Board, Dhaka, Bangladesh.

Brown C. & Durst P. B. 2003. State of Forestry in Asia and the Pacific-2003: Status, Changes and Trends, RAP Publication 2003/22 edn. Asia-Pacific Forestry Commission, Food and Agriculture Organization (FAO) of the United Nations, Bangkok, Thailand.

Brown L. R. 2006. *Plan B 2.0: Rescuing a Planet Under Stress and a Civilization in Trouble*. Earth Policy Institute, New York.

Brown S. 1997. *Estimating Biomass Changes of Tropical Forests: A Primer*, FAO Forestry Paper 134 edn. Food and Agriculture Organization (FAO), Rome, Italy.

Brown S., Gillespie A. J. R., & Lugo A. E. 1989. Biomass estimation methods for tropical forests with applications to forest inventory data. *Forest Science* 35: 881–902.

Brown S., Sathaye J., Cannel M., & Kauppi P. 1996. Management of forests for mitigation of greenhouse gas emissions. In: *Climate Change 1995: Impacts, Adaptations, and Mitigation of Climate Change: Scientific-Technical Analyses* (eds R. T. Watson, M. C. Zinyowera, and R. H. Moss) Cambridge University Press, Cambridge, UK.

Cannell M. G. R. 2003. Carbon sequestration and biomass energy offset: Theoretical, potential and achievable capacities globally, in Europe and the UK. *Biomass and Bioenergy* 24: 97–116.

Chun Y. W. 2002. *Forest and People*. Soomoon Publishing Company, Seoul, The Republic of Korea, 285p.

Dasgupta S., Rahman M. M., Rahman M. L., & Azad A. K. 1990. Agroforestry status in homestead area of Vaskarkhilla FSR site, Kishoregonj. In: *Homestead Plantation and Agroforestry in Bangladesh* (eds M. Z. Abedin, C. K. Lai, and M. O. Ali) pp. 19–54. Bangladesh Agricultural Research Institute (BARI), Joydebpur, Bangladesh.

den Elzen M., Fuglestvedt J., Hohne N., Trudinger C., Lowe J., Matthews B., Romstad B., de Campos C. P., & Andronova N. 2005. Analysing countries' contribution to climate change: Scientific and policy-related choices. *Environmental Science & Policy* 8: 614–636.

Domac J., Richards K., & Risovic S. 2005. Socio-economic drivers in implementing bioenergy projects. *Biomass and Bioenergy* 28: 97–106.

Donahue R. L., Miller R. W., & Shicklana J. C. 1987. *Soils: An Introduction to Soils and Plant growth*, 5th edn. Prentice Hall, New Delhi.

Elauria J. C., Castro M. L. Y., & Racelis D. A. 2003. Sustainable biomass production for energy in the Philippines. *Biomass and Bioenergy* 25: 531–540.

Ellis J., Winkler H., Corfee-Morlot J., & Gagnon-Lebrun F. 2007. CDM: Taking stock and looking forward. *Energy Policy* 35: 15–28.

ESSD (Environmentally and Socially Sustainable Development). 1998. *Greenhouse Gas Assessment Handbook – A Practical Guidance Document for the Assessment of Project-level Greenhouse Gas Emissions*, 64th edn. The World Bank, Washington, D.C., USA.

Ezzati M., Bailis R., Kammen D. M., Hooloway T., Price L., Cifuentes L. A., Barnes B., Chaurey A., & Dhanapala K. N. 2004. Energy management and global health. *Annual Review of Environment and Resources* 25: 383–419.

FAO (Food and Agriculture Organization). 1998. *Asia-Pacific Forestry Sector Outlook Study: Country Report – Bangladesh*, APFSOS/WP/48 edn. FAO Regional Office for Asia and the Pacific, Bangkok, Thailand.

FAO. 2000. *FRA 2000: Forest Resources of Bangladesh*. Country Report. FAO, Rome, Italy.

FAO. 2001. State of the World's Forests 2001. Food and Agriculture Organization of the United Nations, Rome, Italy.

Fearnside P. M. 1999. Forests and global warming mitigation in Brazil: Opportunities in the Brazilian forest sector for responses to global warming under the "clean development mechanism". *Biomass and Bioenergy* 16: 171–189.

Fearnside P. M. 2006. Tropical deforestation and global warming. *Science* 312: 1137.

Gafur A., Karim A., & Khan M. A. A. 1979. Phytosociological studies of hills of the Chittagong University Campus. *Chittagong University Studies, Part II: Science* 3: 11–14.

Gain P. 1998. *The Last Forests of Bangladesh*. Society for Environment and Human Development (SEHD), Dhaka, Bangladesh.

García-Oliva F. & Masera O. R. 2004. Assessment and measurement issues related to soil carbon sequestration in Land-Use, Land-Use Change, and Forestry (LULUCF) projects under the Kyoto Protocol. *Climatic Change* 65: 347–364.

GOB (Government of Bangladesh). 1993. Forestry Master Plan: 3rd Forestry Forum Background Paper, TA No. 1355-BAN edn. Asian Development Bank, Dhaka, Bangladesh.

GOB. 2004. *Bangladesh Economic Review 2004*. Ministry of Finance, Government of the People's Republic of Bangladesh, Dhaka, Bangladesh.

GoRok (Government of the Republic of Korea). 2006. Second National Communication of the Republic of Korea Under the United Nations Framework Convention on Climate Change. Available at http://www.unfccc.int/(Last visited: 29 December 2006).

Gowda M. C., Raghavan G. S. V., Ranganna B., & Barrington S. 1995. Rural waste management in a south Indian village – A case study. *Bioresource Technology* 53: 157–164.

Graham P. 2003. Potential for climate change mitigation through afforestation: An economic analysis of fossil fuel substitution and carbon sequestration benefits. *Agroforestry Systems* 59: 85–95.

Gundimeda H. 2004. How 'sustainable' is the 'sustainable development objective' of CDM in developing countries like India? *Forest Policy and Economics* 6: 329–343.

Guthrie G. A. & Kumareswaran D. K. 2003. Carbon subsidies and optimal forest management. Social Science Research Network (SSRN), New Zealand Institute for the study of competition and regulation. 15p. http://iscr.co.nz/

Hansen E. M., Christensen B. T., Jensen L. S., & Kristensen K. 2004. Carbon sequestration in soil beneath long-term *Miscanthus* plantations as determined by $^{13}C$ abundance. *Biomass and Bioenergy* 26: 97–105.

Healey J. R., Price C., & Tay J. 2000. The cost of carbon retention by reduced impact logging. *Forest Ecology and Management* 139: 237–255.

Hocking D. & Islam K. 1998. Trees on farms in Bangladesh: 5. Growth of top- and root-pruned trees in wetland rice fields and yields of understory crops. *Agroforestry Systems* 39: 101–115.

Hossain M. M. G. 2003. Improved cookstove and biogas programmes in Bangladesh. *Energy for Sustainable Development* 7: 97–100.

Houghton J. 2005. Global warming. *Reports on Progress in Physics* 68: 1343–1403.

Huq S. 2001. Climate change and Bangladesh. *Science* 294: 1617.

Hyman E. L. 1994. Fuel substitution and efficient woodstoves: Are they the answers to the fuelwood supply problem in northern Nigeria? *Environmental Management* 18: 23–32.

Iftekhar M. S. 2006. Forestry in Bangladesh: An overview. *Journal of Forestry* 104: 148–153.

Iftekhar M. S. & Islam M. R. 2004. Managing mangroves in Bangladesh: A strategy analysis. *Journal of Coastal Conservation* 10: 139–146.

IGES (Institute for Global Environmental Strategies). 2010. IGES CDM project database. Available at http://www.iges.or.jp/ (Last visit: 26 February 2010).

IPCC (Intergovernmental Panel on Climate Change). 1992. *Climate Change 1992: The Supplementary Report to the IPCC Scientific Assessment*. Cambridge University Press, Cambridge, UK.

IPCC. 1996. *Revised 1996 IPCC Guidelines for National Greenhouse Gas Inventories: Reference Manual*, Volume 3 edn. Intergovernmental Panel on Climate Change (IPCC), Geneva, Switzerland.

IPCC. 2000. *Land Use, Land-Use Change, and Forestry: A Special Report of the Intergovernmental Panel on Climate Change*. Cambridge University Press, New York.

IPCC. 2001a. Climate Change 2001: Synthesis Report. A Contribution of Working Groups I, II and III to the Third Assessment Report of the Intergovernmental Panel on Climate Change. Cambridge University Press, Cambridge, UK.

IPCC. 2001b. Climate Change 2001: The Scientific Basis. Contribution of Working Group I to the Third Assessment Report of the Intergovernmental Panel on Climate Change. Cambridge University Press, Cambridge, UK.

IPCC. 2001c. *Climate Change 2001: Mitigation. Contribution of Working Group III to the Third Assessment Report of the Intergovernmental Panel on Climate Change*. Cambridge University Press, Cambridge, UK.

IPCC. 2003. Good Practice Guidance for Land Use, Land-use Change and Forestry. Institute for Global Environmental Strategies (IGES), Japan.

Islam K. R., Kamaluddin M., Bhuiyan M. K., & Badruddin A. 1999. Comparative performance of exotic and indigenous forest species for tropical semievergreen degraded forest land reforestation in Chittagong, Bangladesh. *Land Degradation & Development* 10: 241–249.

Ismail R. 1995. An economic evaluation of carbon emission and carbon sequestration for the forestry sector in Malaysia. *Biomass and Bioenergy* 8: 281–292.

Jashimuddin M., Masum K. M., & Salam M. A. 2006. Preference and consumption pattern of biomass fuel in some disregarded villages of Bangladesh. *Biomass and Bioenergy* 30: 446–451.

Jha M. N., Gupta M. K., & Raina A. K. 2001. Carbon sequestration: Forest soil and land use management. *Annals of Forestry* 9: 249–256.

Johns T. C., Gregory J. M., Ingram W. J., Johnson C. E., Jones A., Lowe J. A., Mitchell F. B., Roberts D. L., Sexton M. H., Stevenson D. S., Tett F. B., & Woodage M. J. 2003. Anthropogenic climate change for 1860 to 2100 simulated with the HadCM3 model under updated emissions scenarios. *Climate Dynamics* 20: 583–612.

Kabir M. E. & Webb E. L. 2005. Productivity and suitability analysis of social forestry woodlot species in Dhaka Forest Division, Bangladesh. *Forest Ecology and Management* 212: 243–252.

Kar N. K., Islam S. M. F., Islam S., Abedin M. Z., Musa A. M., & Khair A. B. M. A. 1990. Homestead plantations and agroforestry systems in Barind tract of Bangladesh. In: *Homestead Plantation and Agroforestry in Bangladesh* (eds M. Z. Abedin, C. K. Lai, and M. O. Ali) pp. 19–54. Bangladesh Agricultural Research Institute (BARI), Joydebpur, Bangladesh.

Karjalainen T. 1996. The carbon sequestration potential of unmanaged forest stands in Finland under changing climatic conditions. *Biomass and Bioenergy* 10: 313–329.

Kennett S. A. 2002. National policies for biosphere greenhouse gas management: Issues and opportunities. *Environmental Management* 30: 595–608.

KFS (Korea Forest Service). 2006. Forest distribution: Republic of Korea. Available at http://www.foa.go.kr (Last visited: 29 December 2006).

Khan A. H. M. R., Eusuf M., Prasad K. K., Moerman E., Cox M. G. D. M., Visser A. M. J., & Drisser J. A. J. 1995. The development of improved cooking stove adapted to the conditions in Bangladesh. Final report of collaborative research project between IFRD, BCSIR, Dhaka, Bangladesh and Eindhoven University of Technology, Eindhoven, The Netherlands.

Khan M. S. & Alam M. K. 1996. Homestead Flora of Bangladesh. Bangladesh Agricultural Research Council (BARC); International Development Research Center (IDRC); Swiss Development Cooperation (SDC), Dhaka.

Khan N. A. 1998. Land tenurial dynamics and participatory forestry management in Bangladesh. *Public Administration and Development* 18: 335–347.

Khan N. A. 2001. Regional study on forest policy and institutional reform: Final report of the Bangladesh case study. Asian Development Bank (ADB), Manila, The Philippines.

Kim P. N., Knorr W., & Kim S. 2004. Appropriate measures for conservation of terrestrial carbon stocks-Analysis of trends of forest management in Southeast Asia. *Forest Ecology and Management* 191: 283–299.

Kirschbaum M. U. F. 2001. The role of forests in the global carbon cycle. In: *Criteria and Indicators for Sustainable Forest Management* (eds R. J. Raison, A. Brown, and D. Flinn) pp. 311–339. CABI Publishing, Wallingford, UK.

Koh M. P. & Hoi W. K. 2003. Sustainable biomass production for energy in Malaysia. *Biomass and Bioenergy* 25: 517–529.

Koopmans A. 2005. Biomass energy demand and supply for South and South-East Asia-assessing the resource base. *Biomass and Bioenergy* 28: 133–150.

Kram T., Morita T., Riahi K., Roehrl R. A., Van Rooijen S., Sankovski A., & De Vries B. 2000. Global and regional greenhouse gas emissions scenarios. *Technological Forecasting and Social Change* 63: 335–371.

Kumar B. M. & Nair P. K. R. 2004. The enigma of tropical homegardens. *Agroforestry Systems* 61–62: 135–152.

Lal R. 2004. Soil carbon sequestration impacts on global climate change and food security. *Science* 304: 1623–1627.

Lal R. 2005. Forest soils and carbon sequestration. *Forest Ecology and Management* 220: 242–258.

LeBlanc A. 1999. Issues related to including forestry-based offsets in a GHG emissions trading system. *Environmental Science & Policy* 2: 199–206.

Lee D. K 2003. Forest rehabilitation: Looking east. *FRIM in Focus* October–November–December, 2003. Forest Research Institute of Malaysia (FRIM), Kuala Lumpur, Malaysia.

Lee K. H. 2002. Carbon sequestration of reforestation activities. In: *Restoration of Degraded Forest Ecosystem in Southeast Asia* pp. 25–31. ASEAN-Korea Environmental Cooperation Project (AKECOP).

Lush L., Cleland J., Lee K., & Walt G. 2000. Politics and fertility: A new approach to population policy analysis. *Population Research and Policy Review* 19: 1–28.

Madlener R. & Myles H. 2000. Modelling Socio-Economic Aspects of Bioenergy Systems: A survey prepared for IEA Bioenergy Task 29. Available at http://www.iea-bioenergy-task29.hr/(Last visited: 29 December 2006).

Marland G., Fruit K., & Sedjo R. 2001. Accounting for sequestered carbon: The question of permanence. *Environmental Science & Policy* 4: 259–268.

Marland G. & Schlamadinger B. 1997. Forests for carbon sequestration or fossil fuel substitution? A sensitivity analysis. *Biomass and Bioenergy* 13: 389–397.

Mayaux P., Holmgren P., Achard F., Eva H., Stibig H., & Branthomme A. 2005. Tropical forest cover change in the 1990s and options for future monitoring. *Philosophical Transactions of the Royal Society B: Biological Sciences* 360: 373–384.

Meng F. R., Bourque C. P. A., Oldford S. P., Swift D. E., & Smith H. C. 2003. Combining carbon sequestration objectives with timber management planning. *Mitigation and Adaptation Strategies for Global Change* 8: 371–403.

Miah D., Ahmed R., & Uddin M. B. 2003. Biomass fuel use by the rural households in Chittagong region, Bangladesh. *Biomass and Bioenergy* 24: 277–283.

Miah G., Abedin M. Z., Khair A. B. M. A., Shahidullah M., & Baki A. J. M. A. 1990. Homestead plantation and household fuel situation in Ganges Floodplain of Bangladesh. In: *Homestead Plantation and Agroforestry in Bangladesh* (eds M. Z. Abedin, C. K. Lai, and M. O. Ali) pp. 120–135. Bangladesh Agricultural Research Institute (BARI), Joydebpur, Bangladesh.

Miah M. D. & Hossain M. K. 2002. Tree resources in the floodplain areas of Bangladesh. *Swiss Forestry Journal* 153: 385–391.

Miah M. D., Rahman M. M., & Haque S. M. S. 2001. Carbon assimilation in three-year old mixed plantations in Chittagong region of Bangladesh. *The Chittagong University Journal of Science* 25: 11–16.

Mirza M. M. Q. 2003. Climate change and extreme weather events: Can developing countries adapt? *Climate Policy* 3: 233–248.

Misana S. & Karlsson G. 2001. Generating opportunities: case studies on energy and women. UNDP. http://www.undp.org/

Mishra V., Retherford R. D., & Smith K. R. 2005a. Cooking smoke and tobacco smoke as risk factors for stillbirth. *International Journal of Environmental Health Research* 15: 397–410.

Mishra V., Smith K. R. & Retherford R. D. 2005b. Effects of cooking smoke and environmental tobacco smoke on acute respiratory infections in young Indian children. *Population & Environment* 26: 375–396.

Missfeldt F. & Haites E. 2001. The potential contribution of sinks to meeting Kyoto Protocol commitments. *Environmental Science & Policy* 4: 269–292.

MoEF (Ministry of Environment and Forests). 2005. National Adaptation Programme of Action (NAPA). UNDP and MoEF, Government of the People's Republic of Bangladesh, Dhaka, Bangladesh.

MoERoK (Ministry of Environment of the Republic of Korea). 1997. Republic of Korea: Country profile-Implementation of agenda 21: Review of progress made since the United Nations Conference on Environment and Development, 1992. Available at http://www.un.org/esa/(Last visited: 29 December 2006).

Moura-Costa P. 1996. Tropical forestry practices for carbon sequestration: A review and a case study from Southeast Asia. *Ambio* 25: 279–283.

Myers N. 1996. The world's forests: Problems and potentials. *Environmental Conservation* 23: 156–168.

Niles J. O., Brown S., Pretty J., Ball A. S., & Fay J. 2002. Potential carbon mitigation and income in developing countries from changes in use and management of agricultural and forest lands. *Philosophical Transactions of the Royal Society A: Mathematical, Physical and Engineering Sciences* 360: 1621–1639.

Niroula G. S. & Thapa G. B. 2005. Impacts and causes of land fragmentation, and lessons learned from land consolidation in South Asia. *Land Use Policy* 22: 358–372.

Nordell B. 2003. Thermal pollution causes global warming. *Global and Planetary Change* 38: 305–312.

Olmos S. 2001. Vulnerability and adaptation to climate change: concepts, issues, assessment methods. Climate Change Knowledge Network.

Olsen K. H. & Fenhann J. 2008. Sustainable development benefits of clean development mechanism projects: A new methodology for sustainability assessment based on text analysis of the project design documents submitted for validation. *Energy Policy* 36: 2819–2830.

Osman K. T., Islam M. S., & Haque S. M. S. 1992. Performance of some fast growing trees in the University of Chittagong Campus. *Indian Forester* 118: 858–859.

Osman K. T., Rahman M. M., & Barua P. 2001. Effect of some tree species on soil properties in Chittagong University campus, Bangladesh. *Indian Forester* 127: 431–442.

Ouedraogo B. 2006. Household energy preferences for cooking in urban Ouagadougou, Burkina Faso. *Energy Policy* 34: 3787–3795.

Pearson T., Walker S., & Brown S. 2005. *Sourcebook for Land Use, Land-Use Change and Forestry Projects*. Winrock International, Washington, DC, USA.

Perera K. K. C. K., Rathnasiri P. G., & Sugathapala A. G. T. 2003. Sustainable biomass production for energy in Sri Lanka. *Biomass and Bioenergy* 25: 541–556.

Petersen L. 1996. *Soil Analytical Methods*. Soil Resource Development Institute (SRDI), Dhaka, Bangladesh.

Pussinen A., Karjalainen T., Kellomaki S., & Makipaa R. 1997. Potential contribution of the forest sector to carbon sequestration in Finland. *Biomass and Bioenergy* 13: 377–387.

Qiu D., Gu S., Catania P., & Huang K. 1996. Diffusion of improved biomass stoves in China. *Energy Policy* 24: 463–469.

Quay P. 2002. Climate change: Ups and downs of $CO_2$ uptake. *Science* 298: 2344.

Ramachandra T. V., Joshi N. V., & Subramanian D. K. 2000. Present and prospective role of bioenergy in regional energy system. *Renewable and Sustainable Energy Reviews* 4: 375–430.

Ravindranath N. H., Balachandra P., Dasappa S., & Usha Rao K. 2006. Bioenergy technologies for carbon abatement. *Biomass and Bioenergy* 30: 826–837.

Ravindranath N. H. & Somashekhar B. S. 1995. Potential and economics of forestry options for carbon sequestration in India. *Biomass and Bioenergy* 8: 323–336.

Reddy B. S. & Balachandra P. 2006. Dynamics of technology shifts in the household sector-implications for clean development mechanism. *Energy Policy* 34: 2586–2599.

Roper J. 2001. *Forestry Issues: Tropical Forests and Climate Change*. CFAN (CIDA Forestry Advisers Network), Canadian Institute of Development Agency, Quebec, Canada.

Rosenbaum K. L., Schoene D., & Mekouar A. 2004. Climate Change and the Forest Sector: Possible National and Subnational Legislation, FAO Forestry Paper 144 edn. Food and Agriculture Organization (FAO) of the United Nations, Rome, Italy.

Rubab S. & Kandpal T. C. 1996. Biofuel mix for cooking in rural areas: Implications for financial viability of improved cookstoves. *Bioresource Technology* 56: 169–178.

Sajjakulnukit B. & Verapong P. 2003. Sustainable biomass production for energy in Thailand. *Biomass and Bioenergy* 25: 557–570.

Salam M. A. & Noguchi T. 2005. Impact of human activities on carbon dioxide ($CO_2$) emissions: A statistical analysis. *The Environmentalist* 25: 19–30.

Salam M. A., Noguchi T., & Koike M. 1999. The causes of forest cover loss in the hill forests in Bangladesh. *Geojournal* 47: 539–549.

Salam M. A., Noguchi T., & Koike M. 2005. Factors influencing the sustained participation of farmers in participatory forestry: A case study in central Sal forests in Bangladesh. *Journal of Environmental Management* 74: 43–51.

Sattar M. A., Bhattacharjee D. K., & Kabir M. F. 1999. Physical and Mechanical Properties and Uses of Timbers of Bangladesh. Seasoning and Timber Physics Division, Bangladesh Forest Research Institute (BFRI), Chittagong, Bangladesh.

Schulze E. D., Wirth C., & Heimann M. 2000. Climate change: Managing forests after Kyoto. *Science* 289: 2058–2059.

SEHD (Society for Environment and Human Development). 1998. *Bangladesh Environment Facing the $21^{st}$ Century*. Society for Environment and Human Development, Dhaka, Bangladesh.

Sen B. 2003. Drivers of escape and descent: Changing household fortunes in rural Bangladesh. *World Development* 31: 513–534.

SESRTCIC (Statistical, Economic and Social Research and Training Centre for Islamic Countries). 2006. SESRTCIC Statistical Database. Available at http://www.sesrtcic.org/statistics/(Last visited: 29 December 2006).

Shin M. Y., Miah M. D., Sadeq M., & Lee K. H. 2004. Homestead agroforestry products and their utilization in the Old Brahmaputra floodplain area of Bangladesh. *Journal of Korean Forest Society* 93: 373–382.

Silveira S. 2005. Promoting bioenergy through the clean development mechanism. *Biomass and Bioenergy* 28: 107–117.

Sims R. E. H. 2003. *Bioenergy Options for a Cleaner Environment: In Developed and Developing Countries*, 1st edn. Elsevier Ltd., Oxford, UK.

Smith K. R. 2002a. Indoor air pollution in developing countries: Recommendations for research. *Indoor Air* 12: 198–207.

Smith J. 2002b. Afforestation and reforestation in the clean development mechanism of the Kyoto Protocol: Implications for forests and forest people. *International Journal of Global Environmental Issues* 2: 322–343.

Smith J. & Scherr S. J. 2003. Capturing the value of forest carbon for local livelihoods. *World Development* 31: 2143–2160.

Smith K. R., Shuhua G., Kun H., & Daxiong Q. 1993. One hundred million improved cookstoves in China: How was it done? *World Development* 21: 941–961.

Smith K. R., Uma R., Kishore V. V. N., Zhang J., Joshi V., & Khalil M. A. K. 2000. Greenhouse implications of household stoves: An Analysis for India. *Annual Review of Energy and the Environment* 25: 741–763.

Sohngen B. & Mendelsohn R. 2003. An optimal control model of forest carbon sequestration. *American Journal of Agricultural Economics* 85: 448–457.

SSS (Soil Survey Staff). 1979. Detail Soil Survey of Chittagong University Campus, Chittagong, Bangladesh. Soil Resources Development Institute (SRDI), Dhaka, Bangladesh.

Stebich M., Bruchmann C., Kulbe T., & Negendank J. F. W. 2005. Vegetation history, human impact and climate change during the last 700 years recorded in annually laminated sediments of Lac Pavin, France. *Review of Palaeobotany and Palynology* 133: 115–133.

Streets D. G. & Waldhoff S. T. 1999. Greenhouse-gas emissions from biofuel combustion in Asia. *Energy* 24: 841–855.

Sudha P. & Ravindranath N. H. 1999. Land availability and biomass production potential in India. *Biomass and Bioenergy* 16: 207–221.

Sudha P., Somashekhar H. I., Rao S., & Ravindranath N. H. 2003. Sustainable biomass production for energy in India. *Biomass and Bioenergy* 25: 501–515.

Takahashi T. 2004. The fate of industrial carbon dioxide. *Science* 305: 352–353.

Tipper R. & de Jong B. H. 1998. Quantification and regulation of carbon offsets from forestry: Comparison of alternative methodologies, with special reference to Chiapas, Mexico. *Commonwealth Forestry Review* 77: 219–228.

Tschakert P. 2004. Carbon for farmers: Assessing the potential for soil carbon sequestration in the Old Peanut Basin of Senegal. *Climatic Change* 67: 273–290.

UNEP (United Nations Environment Program). 1994. *Environmental Data Report 1993–94*. UNEP, London, UK.

UNEP. 2001. Bangladesh: State of the Environment 2001. Available at http://www.rrcap.unep.org/reports/(Last visited: 29 December 2006).

UNFCCC (United Nations Framework Convention on Climate Change). 2006a. Kyoto Protocol: Status of Ratification. Available at: http://www.unfccc.int/(Last visited: 29 December 2006).

UNFCCC. 2006b. CDM: Guidance to the EB (COP9) – Decision 18/CP.9 (FCCC/CP/2003/6/Add.2). Available at: http://www.unfccc.int/(Last visited: 29 December 2006).

UNFCCC. 2006c. Parties & Observers: Non-Annex I-Bangladesh. Available at: http://www.unfccc.int/(Last visited: 29 December 2006).

UNFCCC. 2006d. Parties & Observers: Non-Annex I-Republic of Korea. Available at: http://www.unfccc.int/(Last visited: 29 December 2006).

USCB (U.S. Bureau of the Census). 2006. International Data Base (IDB): Population Division. Available at: http://www.census.gov/(Last visited: 29 December 2006).

Vincens A., Williamson D., Thevenon F., Taieb M., Buchet G., Decobert M., & Thouveny N. 2003. Pollen-based vegetation changes in southern Tanzania during the last 4200 years: Climate change and/or human impact. *Palaeogeography, Palaeoclimatology, Palaeoecology* 198: 321–334.

Võhringer F. 2004. Forest conservation and the clean development mechanism: Lessons from the Costa Rican protected areas project. *Mitigation and Adaptation Strategies for Global Change* 9: 217–240.

WB (The World Bank). 1998. World Development Indicators 1998. Oxford University Press, Washington, D.C., USA.

WEC (World Energy Council). 2000. Renewable Energy in South Asia: Status and Prospects. World Energy Council (WEC), London, UK.

Zafarullah H. & Siddiquee N. A. 2001. Dissecting public sector corruption in Bangladesh: Issues and problems of control. *Public Organization Review: A Global Journal* 1: 465–486.

# Subject Index